T0123152

BREAKING AWAY from the MATH BOOK II

More Creative Projects for Grades K–8

BREAKING AWAY from the MATH BOOK II

More Creative Projects for Grades K–8

PATRICIA BAGGETT
ANDRZEJ EHRENFEUCHT

SCARECROWEDUCATION
LANHAM, MARYLAND • TORONTO • OXFORD

Published in the United States of America
by ScarecrowEducation
An imprint of The Rowman & Littlefield Publishing Group, Inc.
4501 Forbes Boulevard, Suite 200, Lanham, Maryland 20706
www.scarecroweducation.com

PO Box 317
Oxford
OX2 9RU, UK

The original edition of *Breaking Away from the Math Book II* was published in 1998 by Technomic Publishing Company, Inc.

Copyright © 1998 by Technomic Publishing Company, Inc.
First ScarecrowEducation paperback edition 2004

All rights reserved. No part of this publication may be reproduced, stored in a retrieval system, or transmitted in any form or by any means, electronic, mechanical, photocopying, recording, or otherwise, without the prior permission of the publisher.

Baggett, Patricia.
  Breaking away from the math book II : more creative projects for grades k–8 / Patricia Baggett, Andrzej Ehrenfeucht.
      p. cm.
  Originally published: Lancaster, Pa. : Technomic Pub. Co., c1998.
  ISBN 1-57886-160-8 (pbk. : alk. paper)
  1. Mathematics—Study and teaching (Elementary) I. Ehrenfeucht, Andrzej. II. Title.
  QA135.6.B339 2004
  372.7—dc22
                                                                            2004013023

⊖™The paper used in this publication meets the minimum requirements of American National Standard for Information Sciences—Permanence of Paper for Printed Library Materials, ANSI/NISO Z39.48-1992.
Manufactured in the United States of America.

In memory of David Eugene Smith (1860–1944)

TABLE OF CONTENTS

Part B. Three-dimensional Constructions

SECTION VII. PYTHAGOREAN THEOREM

SECTION VIII. MISCELLANEOUS

SECTION IX. LEARNING HOW TO USE A CALCULATOR

The National Council of Teachers of Mathematics recommended the use of calculators in 1989. While calculators are more readily available in classrooms today than they were a few years ago, the recommendation is rarely followed in elementary grades. One of the reasons is that there is a lack of appropriate classroom materials. This book provides well tested lessons in which calculators play the role of tools helping to solve mathematical problems. Children who learn how to combine mental calculations with the use of calculators become better problem solvers, because they can concentrate on the problem itself, and not on mechanical details of the process of computation.

We have also observed positive changes in children's attitudes toward mathematics. Many children develop a fear of mathematics early in their schooling. "I'll get it all wrong again," they think. They have a lack of confidence in their own solutions, because of a high rate and a persistence of arithmetic errors. When children are properly taught how to use calculators, arithmetic errors practically disappear from their solutions. This gives them a well-justified confidence in their own work, and it removes the most common reason for disliking mathematics, the fear of being wrong.

This book is a continuation of *Breaking Away from the Math Book: Creative Projects for Grades K–6*, but it contains a number of more advanced lessons appropriate for grades 7 and 8. The last section, Chapter 61, is a tutorial on how to use a four-operation calculator.

Since the publication in 1995 of *Breaking Away from the Math Book: Creative Projects for Grades K–6*, our testing ground for the lessons has changed. Beginning in fall 1995 we have been working with K–8 teachers in the Las Cruces, New Mexico, Public Schools. With the help of Karin Matray, Director of Staff Development for the Las Cruces Public Schools, together with Feliciano Mendosa, resource teacher for the Las Cruces Teachers' Center, and Dr. Martha Cole, Associate Superintendent for Instruction, LCPS, a partnership has been arranged with some K–8 teachers and the Department of Mathematical Sciences at New Mexico State University. Teachers receive graduate credit for a special topics course, Math 501, and attend an experimental section of Math 111/112 (Fundamentals of Elementary Mathematics I and II), which is required for undergraduate elementary education majors (preservice teachers). The inservice teachers allow the preservice teachers to observe in their classrooms, coteach, and finally teach alone. This has become an excellent way to test lessons in this book: We try them in the university class, and then teachers and preservice teachers try them with children in classrooms. Teachers have also allowed Professor Baggett to try lessons in their classrooms on a regular basis. Children, together with inservice and preservice teachers, have given us enthusiastic feedback about the lessons. We are grateful to the administrators, teachers, and children who have supported our efforts.

The teachers who have participated thus far, their schools, and principals, are:

- Alameda Elementary School, Pete Tierney, Principal; Joan Hirsh, 2nd grade; Shani Magee, 4th grade; Sandra Nakamura, 1st grade
- Central Elementary School, Barbara Morrison, Principal; John Troy, 4th grade; Beverly Whygles, 4th, 5th, and special education
- Davenport Elementary, Cantillo, TX, Sylvia Gonzalez, Principal; Amy Craig, K, K–3 math lead teacher
- East Picacho Elementary School, Jerry Melder, Principal; Michelle Boudreau, 2nd grade; Lisa Girard, 5th grade; Amy Hall, 2nd grade; Elizabeth Johnson, kindergarten
- Fairacres Elementary School, Theresa Jaramillo-Jones, Principal; Judy Foster, 1st grade; Sherri Ploss, 4th grade; Peggy Williams, 4th grade
- Hermosa Heights Elementary School, Irma Glover, Principal; Margaret Calderon, 3rd grade
- Hillrise Elementary School, Clyde Walton, principal; Jean Bishop, 5th grade
- Jornada Elementary School, Phil Allen, Principal; Linda Tierney, grades K–2; Win

Krause, grades K–2; Renee Mullis, grades 3, 4, 5 and special education; Joyce Watson, 2nd grade; Hank Hopkins, 3rd grade; Shirley O'Brian, grades K–1–2

- Lynn Middle School, Al Ioerger, Principal; Patsy Padilla, 6th grade
- Loma Heights Elementary School, Daniel Miller, Principal; Yvette Ramirez-Springler, 2nd grade; Amy Stevens, 2nd grade; Pat Waugh, 1st and 2nd grades; Sam Hoffman, 1st and 2nd grades
- MacArthur Elementary School, Carlos Carrillo, Principal; Linda Callender, 4th grade; Filo Rigales, 1st grade
- Mesilla Alternative School, Eric Cress, Principal; Clark Keith, 8th and 9th grades
- Mesilla Elementary School, Calle del Sur, Mesilla, Liz Marrufo, Principal; Elaine Dinger, 1st grade; Teresa Gamboa, 4th grade
- Mesilla Park Elementary School, Nancy Ayers, Principal; Beverly Shock, kindergarten
- Sierra Middle School, Jean de la Peña, Principal; Anna Suggs, 6th grade; Steven Dubrava, 5th grade; Terrie Hansen, 8th grade
- Sunrise Elementary School, Richard Schriver, Principal; Marcia Anderson, 1st grade; Polly Hirtzel-Ward, 1st grade; Kelley Bagley, 2nd grade; Judy Jacobi, 5th grade; Priscilla Mohrhauser, 1st grade; Netta Pogue, 4th grade
- Tombaugh Elementary School, Chris Milyard, Principal; Brian Culberson, kindergarten; Sarah Priestley, kindergarten; Michelle Varoz, special education; Kathleen Wittman, 2nd through 5th grades
- University Hills Elementary School, Vincent Rivera, Principal; Cindy James, 1st grade; Shirley Stevens, kindergarten; Judy Walker, 3rd grade
- Valley View Elementary, Ellie Chalekian, Principal; Denise Lucht, 3rd grade; Joe Zuniga, 5th grade
- Vista Middle School, Olivia Ogas, Principal; Mary Patterson, 6th grade; Dana Domme, 7th grade; Pamela Merrick, 7th grade; Gina Rivera, 8th grade
- Washington Elementary, Robert Sanchez, Principal; Linda Corona, 1st grade; Bea Acosta, 4th grade; Arthur Padilla, 5th grade
- White Sands Elementary School, James Telles, Principal; Cathy Boeker, 4th grade
- Zia Middle School, Thomas Hayes, Principal; Diana Lawton, 6th grade; Donna Parish, 7th grade

We thank the National Science Foundation for sponsoring the development of the curricular materials from 1994 until 1997 under grant number ESI-9596198. We also thank the Las Cruces Teachers' Center for sponsoring the NMSU/Las Cruces Public Schools partnership program in spring 1996, and the New Mexico Commission on Higher Education for sponsoring it from fall 1996 through spring 1998. Finally, we thank Leesa Mandlman and Ali Ahmad for preparing illustrations for the book.

Learning mathematics can be challenging, interesting, and rewarding. But spending hours on mindless practice can dampen all but the most persistent spirits. This book contains 60 classroom-tested units. They provide challenging mathematical material for grades kindergarten through eight. They can be used by teachers at school or by parents at home. They connect mathematics to other activities which interest children, and show that mathematics is a powerful problem solving tool.

Solving problems is a very different activity than giving answers to dreaded word problems. Therefore, children should be provided with good quality tools, such as rulers, protractors, compasses, and scissors. Let them work with a variety of objects. If a lesson calls for rocks, do not replace them with Unifix cubes!

Calculators are a must. In early grades, use simple four-operation calculators. More complex calculators, whose use may require knowledge the children will not get until much later, may slow their progress and discourage their effort to understand mathematics.

The units are in three different formats. One, shown for example in *What is the density of a small rock?* is a brief lesson plan. A second, shown for example in *Squirrels' race,* is a longer version, a narrative describing how the lesson was taught in a classroom. A third type is a discussion of a mathematical topic, such as *The concept of infinity,* or *One third.* These units lead to lively discussions in classrooms.

The lessons in this book are for all learners, slow, fast, and in between. Remember, slow learners only require more time; they do not require easier or watered down tasks!

COUNTING AND NUMBERS

CHAPTER 1

Counting Stones by Grouping and Using the Memory Key

This lesson was taught in two first grade classes.

PROPS

Each child should have a cup of stones (40 or 50 maximum); a paper plate to dump the stones onto; and a calculator. The teacher needs an overhead calculator and her own cup of stones.

THE LESSON

Each child was given a cup of stones and allowed to dump them out onto his or her plate. The stones we used came from Lake Superior; they are smooth and come in many colors, from white to yellow to orange to beige to brown to gray to black, and in many sizes and granularities. The children liked just to look at them, to sort them by size or by color, and to move them around, making patterns and pictures on their plates. This free exploration yielded lots of comments and questions by the children. "Where did they come from?" "What makes them so smooth?" "How do some of them get their stripes?" In one classroom I also passed out magnifying glasses, to allow the children to take a closer look at the stones. It would be easy to turn the beginning of this lesson into one about science, to name the minerals (quartz, granite, etc.), and to talk about the origins of the rocks.

After the children had played a while, I asked, "How many stones do you think you have in your cup? How do you think you could find out?"

"We could count them!" replied the children.

"Sure, we could," I replied. "I want to show you a neat way to use your calculator to help you count them. Do you know about the memory in your calculator?" Some children knew about it, but most didn't. "It's a place in the calculator where you can put a number. You know the memory has a number in it when you see an 'M' on the display. And you can look and see what the number in memory is by pressing [MRC], which means 'memory recall'. Let me show you how I am going to count my stones."

I poured my stones out onto the overhead projector, and I turned on my overhead calculator by pressing [ON/C]. "I am going to count my stones one by one. First I am going to put 1 on my calculator display by pressing [1]. Now, one by one, I am going

ILLUSTRATION 1. Counting stones by grouping.

to pick up my stones. Every time I put one in my cup, I am going to press [M+] (memory plus). Watch this."

I picked up a stone and dropped it in my cup, and pressed [M+].

I picked up another, dropped it into the cup, and pressed [M+] again. "Now how many stones do I have in my cup?" I asked.

"Two," replied several.

"Well, then what should I have in memory?"

"Two?" they asked.

"Let's look and see. What do I press to look into memory?"

"MRC!" some replied. I pressed [MRC], and 2 appeared on the display. The children thought this was pretty cool.

"Now I want to go on with my counting. So I'll put 1 on my display again," (I pressed [1]), "And now I'll put another stone in the cup and press [M+]. Do you follow what I am doing? After I put [1] on the display, every time I put a stone in the cup, I press [M+]. That's all there is to it." (As I talked, one by one I kept putting the stones in the cup, and pressing [M+] each time.) "Here's my last one," I said, and I put it in the cup and pressed [M+]. "How many stones do you think I have?"

"Press [MRC] to see!" said Justin. I pressed [MRC], and 42 appeared on the display.

"You've got 42 stones," said Lucy.

"Okay, can you use your calculator to count your stones now?" I asked. "First, after you have your stones dumped onto your plate, turn on your calculator and put 1 on the display. Now, one by one, put a stone in the cup, and press [M+]. Keep doing it until all your stones are in your cup. Then press [MRC] to see how many you have."

The children found this activity engrossing. The classroom was filled with the sounds of stones being dropped into cups. Pretty soon, I heard utterances of surprise from throughout the room.

"I have 49 stones!"

"I have 47!"

"I have 50!"

I walked around to be sure that each child was able to count by using the [M+] key, and that each child had a final total, after all stones were put in his or her cup. "Now, can you remember the number of stones you have?" I asked. "I want to show you another way to count them. First, we need to clear the memory of the calculator. To do that, press [MRC][MRC]. Do you see that the 'M' on your display has gone away? Now your memory is empty." I pressed [MRC][MRC] on the overhead calculator to clear its memory. "Next, pour your stones back onto your plate. Now let me show you a new way to count." I poured my stones back out onto the overhead. "I'll pick up two stones this time, and put them in my cup. And I'll put two on my calculator display and press [M+]." (I put two stones in my cup and pressed [2][M+].) "Now I'll pick up a group of 3 stones, put them in my cup, and press [3][M+]." (I did so.) "Now I'll pick up 2 again, and press [2][M+]. Do you see what I am doing? I pick up some number of stones, whatever I want, and put them in my cup. Then I put that number on the display and press [M+]." I continued to do this until all my stones were in the cup. "How many stones do you think I have?"

"You should have what you had before!" said several.

"How many was that? Does anyone remember?" I asked.

"42!" said Suellen and Bart.

"Let's press [MRC] and see." Sure enough, when I pressed [MRC], 42 showed on the display.

The children thought this was rather surprising. "Now try it yourself. Remember what we are doing. First, be sure your memory is clear, by pressing [MRC][MRC]. Then make groups of as many stones as you want; put them in your cup, and put the number on the display and press [M+]. When you're done, press [MRC] to see how many you have altogether."

I walked around to be sure the children had cleared their calculators and were starting over. They demonstrated several different strategies as they counted. Some counted by twos (as we had counted by ones before); they did not reenter [2] each time, but simply pressed [M+]. Of course, at the end, if only 1 stone was left, they had to enter [1][M+].

Others counted by tens. And others counted by different numbers, as I had demonstrated.

"When you are done, press [MRC] to see your total," I said. "Is it the same as the number you had before?" Most children indicated that they got the same number.

"Well, let's count our stones one more time," I said. "First, clear memory by pressing [MRC][MRC]. Then dump your stones back out onto your plate. This time I'm going to sort my stones by size." I made 4 different groups of stones, very small, small, medium, and large. "Now I'm going to count the number in each group, and put it into the calculator's memory." I counted 5 in the very small group, 10 in the small group, 20 in the medium group, and 7 in the large group. I pressed [5][M+][10][M+][20][M+][7][M+] "How many do you think I have altogether?"

"42!" they replied.

"Let's see." I pressed [MRC], and 42 appeared. "Now you sort your stones into groups, count the number in each group, and put each number into memory. Then press [MRC] to see the total." I walked around to help children with this task. "Did you get the same number you got before?" I asked.

The children had lots of questions about the sorting task. "How do I know if a stone is small or medium? Does it matter how many stones are in each group? How large is large?" But eventually every child was able to do the task, and eventually they were no longer surprised that, no matter how they grouped their stones, the total number was always the same.

It was time to stop, so I collected the cups of stones, plates, and calculators. If time had permitted, I would have collected each child's stones in a common container and kept a running total, so see how many stones the children had in all.

There is no hidden agenda in this lesson. It is meant to show that there are many ways to count by grouping, and each way gives the same number. At first, this fact is not obvious to young children.

CHAPTER 2

Horse Trader

THE LESSON

This lesson has been taught in kindergarten and first grade. Teachers have been very inventive in using it. For example, one kindergarten teacher changed "horse trader" to "pumpkin collector" and taught the lesson near Halloween. Children take great delight in pretending that the stones they are using are horses (or pumpkins, etc.!). I used an overhead transparency with the puzzle on it (see end of lesson), and in first grade I gave each child a copy of the overhead transparency, so they could follow along at their desks. I drew a circle on the overhead, to indicate a corral. The children first read the puzzle line by line, and then made guesses about how many horses the trader would have when he left town.

(1) Talk about the role of horses in the past.
(2) Here is a puzzle.

A horse trader came into a town on a horse.
The first day he bought two horses, and then he sold one.
The next day he bought three horses and then sold two.
Later he bought four horses and sold three.
Finally he bought five horses, sold four, and left town.
How many horses did he have when he left town?

(3) Some solutions

(a) First solution (with manipulatives)

Each child should have about 10 tokens which can stand for "horses" and a small paper plate which is a corral where the horse trader keeps his horses. The story should be told again, and as it is told children put the tokens on the plate, or take them off the plate.
A horse trader came into town on a horse.
(Put one token on the plate. The horse trader has one horse.)
The first day he bought two horses
(Put two tokens on the plate. The horse trader has three horses.)
and then he sold one.
(Take a token from the plate. The horse trader has two horses.)
The next day he bought three horses
(Put three tokens on the plate. The horse trader has five horses.)
and then he sold two.

ILLUSTRATION 2. Just now the horse trader has four horses.

(Take two tokens from the plate. The horse trader has three horses.)
Later he bought four horses
(Put four tokens on the plate. The horse trader has seven horses.)
and sold three.
(Take three tokens from the plate. The horse trader has four horses.)
Finally he bought five horses,
(Put five tokens on the plate. The horse trader has nine horses.)
and he sold four,
(Take four tokens from the plate. The horse trader has five horses.)
and left town.
The horse trader left with five horses.

Comment. The result of each action can be discussed, but it would be good if the story and actions developed a rhythmic pattern.

(b) Second solution (with a calculator)

Now let the numbers be horses, and the calculator's memory be the place where they are kept. (Clear the calculator.)

[1][M+]	A horse trader into town on a horse.
[2][M+]	The first day he bought two horses
[1][M−]	and then he sold one.
[3][M+]	The next day he bought three horses
[MRC]	display: 5 (See how many horses he has now.)
[2][M−]	and then he sold two.
[4][M+]	Later he bought four horses
[3][M−]	and sold three.
[5][M+]	Finally he bought five horses,

[4][M−]	sold four,
[MRC]	display: 5 and left town with five horses.

(c) Other solutions

Another calculator solution

[1]
[+][2][−][1]
[+][3][−][2]
[+][4][−][3]
[+][5][−][4][=]

Mental solution

Notice that each time he bought one more horse than he sold,
$2 - 1 = 1$, $3 - 2 = 1$, $4 - 3 = 1$, and $5 - 4 = 1$.
So he was getting one more horse after each transaction.
Therefore,

He came on	one horse.
After the first day he had	two horses.
The next day he had	three horses.
Later he had	four horses.
Finally he had	five horses,

and he left town with them.

(4) Modify this story, or create another story with a similar pattern.

A puzzle

A horse trader came into a town on a horse.

The first day he bought two horses,
                                 and then he sold one.

The next day he bought three horses
                                 and then he sold two.

Later he bought four horses
                                 and then he sold three.

Finally he bought five horses,
                                 sold four, and left town.

How many horses did he have when he left town?

CHAPTER 3

Jar of Stones

This lesson was taught in a first grade class.

PREREQUISITES

There are none, except being able to count to about 25. It is helpful if children have used the [M+] key on the calculator before, but use of [M+] can be taught during the lesson.

PROPS

The teacher should find beforehand a clear plastic jar (with lid) and should fill it with stones of various colors, shapes, and sizes. She should pack the jar as full as possible (bang it on a table to shake down the stones) and screw on the lid. Then it will not rattle when shaken. The teacher should also have an overhead calculator and a container of water. Each child needs a calculator and (eventually) a cup of stones from the jar.

NOTES

(1) The stones will obviously vary in size, but the size differences should not be huge.
(2) The teacher does not need to know beforehand how many stones are in the jar.
(3) I used stones collected on a trip to the Upper Peninsula of Michigan, along the shore of Lake Superior. They have smooth (not sharp) surfaces and edges, and come in a large array of colors: black, dark gray, red, brown, yellow, off-white, and various multi-colored patterns: speckled, striped, etc. These stones feel good when touched, and their wonderful colors become even more brilliant when they are put in water (as will be explained at the end of the lesson).

THE LESSON

I walked into the classroom with a jar full of stones. I walked among the children's desks, putting the jar down here and there for them to examine. "I want to know something," I said. "How many stones are in this jar?" The kids were definitely curious.

"A million!" said one.

"Forty!" said another.

"No way!" said another. "A thousand!"

ILLUSTRATION 3. How many stones are in the jar?

"A million a million!"

"Seventy-three!"

"Five hundred!"

"Eight hundred and seventy nine!"

The guesses went on and on. I wrote them all on the board. (I wrote "a million a million" for the fourth guess!)

"Well," I said, "How are we going to find out how many there really are?"

There was large agreement here—"We need to count them" was spoken almost in unison by the whole class.

"Okay, how are we going to count them?" I asked.

"One, two, three, . . . " said Latisha.

"That would take a long time," said Brad.

"Yes, it would take a long time for one person," I said. "Any other ideas?"

"You could give a handful of stones to each person in the class, and we could each count our handful," said Angela.

"That's a great idea," I said. "And then what would we need to do?" The class was rather baffled here. "Well, if we pass out all the stones to kids in the class, and we know how many stones each kid has, then how will we know how many stones were in the jar? How many altogether?"

Finally, Dennis said, "We can plus them!"

"Right!" I said. "Would you each like a cup of stones to count?" The children were eager to get their stones. "How many kids are here today?" I asked. "Twenty-two," said Shareen.

I set out 22 cups and poured stones into each, until the stones were more or less equally distributed. "Here's what I suggest," I said. "I'll pass these out. When you count your stones, you can put the number of stones you have onto your calculator display. Then, one by one, you can come up and pour your cup of stones back into the jar, and enter your number of stones into the overhead calculator, and hit the [M+] key. Do you remember what the [M+] key does?" "Yes," said some. "No," said others.

"Well, we will see what it does when we start using it," I said. "The [M+] key will let us keep track of how many stones we have in the jar."

I passed out the cups of stones. This caused great excitement. "Dump your stones out carefully and gently on your desk," I said. "You can just look at them awhile before you need to count them."

Children poured their stones out onto their desks. (A note to teachers: You might use large paper plates here as containers for the stones, but I prefer the freedom of an open space, even if some stones will invariably end up on the floor.) There was a lot of commotion, and statements of awe: "Wow, this one sparkles!" "I have a striped one!" "I can see a face in this one!" "This one is so flat!"

I encouraged the children to sort their stones into groups, by size or color or however they wanted. After a few minutes, I said, "Okay, now will you please count your stones and put the number that you have onto the calculator display? Count carefully, and if you think you might have made a mistake, count again and see if you get the same number. When you're pretty sure you have the right number, that's the number to put on your display."

Children counted, and sometimes they recounted. I did not check their individual counting. But I could see that the number of stones varied between about 17 and 29.

"Do you have your number of stones on your calculator display?" I asked.

"Yes," replied the children.

"Okay then, put your stones back in your cup. Who would like to bring up your cup first?"

Stan volunteered. "I have 19 stones," he announced.

"Okay," I said. "Pour them into the jar, and punch [19][M+] into the overhead calculator." Stan poured his stones into the jar. They did not even cover the bottom! But they made a loud noise as he poured. He then punched [19][M+]. An "M" appeared on the display, along with the number 19.

"Stan, do you know what the M means?" I asked.

"No," he said. Alma said it meant 19 was in the calculator's memory. "We learned about the memory basket in the calculator before," she said.

"That's right," I said. "Nineteen is in memory." I thanked Stan and asked who would like to come up next.

Tonya volunteered. "I have 23 stones," she said. She dumped them into the jar and pressed [23][M+].

"How many stones are in the jar now?" I asked. No one was sure.

"We can look," said Alma.

"How do we look?" I asked.

"Punch [MRC]," she said.

"Right," I said. I pressed [MRC], and 42 appeared on the display.

"Wow!" said several.

"Well," I said, "This is kind of interesting. We already have 42 stones in the jar. Let's go over here to our list of guesses. Are there any guesses here that we already know are not correct?" This question caused some thinking.

Finally Pamela said, "We have 42, so we know 40 is too small."

"Great," I said. And I put a slash through the number 40 on the board.

"Who's next?" I asked.

Samuel volunteered. "I have 27 stones," he said. He poured his stones into the jar and pressed [27][M+] on the overhead calculator.

"Shall we look and see how many stones we have now?" I asked. Some wanted to look, and others didn't. "Okay," I said. "Let's vote. How many want to look now? Raise your hand." (I counted.) "And how many want to wait?" (I counted.) "Most of you want to wait, so let's wait."

We continued in this fashion. One by one, the children came up and dumped their stones into the jar and entered their number of stones onto the calculator display and pressed [M+]. About every second time we looked into memory to see the total, and we went to the list of guesses and decided which ones needed to be crossed off. The children were quite engrossed in this activity. Some told me I could already cross off a million (and a million a million), because there was no way we were going to have so many.

When it was James's turn, he came up and announced, "I have 27 stones." But on his calculator was the number seventy-two.

I asked him, "How do you put the number 27 into your calculator?"

"I press [2] and then [7]," he said.

"Okay," I said. "The [2] goes first, and then the [7]. Try it." He did it correctly on the overhead calculator.

Luellen said she had 22 stones, but she had pressed 55 on her calculator. I gently helped her to correct her error on the overhead. "It's difficult to tell a 2 and a 5 apart on the calculator," I said. "One is just the other one written backwards."

Often we looked into memory to see how many stones we had, and then a child came up and entered his or her number into the calculator and hit [M+]. I would ask the kids to try to figure out (using their "calculators with hair on them," namely, their own heads!) how many were now in memory. Some children could figure this out, especially when the number to be added was 20.

The jar was nearly full (there were 436 stones in it), and there were only two children who had not put in their stones. So I asked, "How many stones do you think we will have when we are done?"

"Not 1000," said several.

"Why not?" I asked.

"Because there are only two more people, and no way will they have enough to make a thousand," said Brad.

"I think you're right," I said, and I crossed off 1000 from the list. "Will we have 500?" I asked. No one was sure. "Well, if we would have 500, then how many more stones would we need? Can anyone figure that out?" This question caused some consternation. "How can we figure it out?" I asked.

Finally Latisha said, "We can take away." But she wasn't sure what to take away. I wrote 500 on the board, with a big minus sign below it. "What should I take away?" I asked. "How many stones do we have in the jar now?"

"436," replied several.

"I know," said Angela. "We take away 436."

"Right," I said. "Will you try it on your calculators for me?" I wrote on the board [500][−][436][=]. "64," came the reply. "We would need 64 more stones."

"All right, let's see how many stones we really have. Let's see, we have Paula's stones and Dennis's stones. Who wants to come up first?"

Dennis volunteered. "I have 21 stones," he said. He entered [21][M+].

"How many do we have now?" I asked. No one was able to say, so I hit [MRC], and the number 457 showed on the display. There were a few ooohs and aaahs. "Do you think we will hit 500?" I asked. Most kids thought we would not. "Paula, will we hit 500?" I asked. Paula said she did not want to tell! "Well, how many will Paula have to have, if we are to get 500 exactly?" I asked. "We can take away again," said several. "Take away what from what?" I asked.

"[500][–][457][=]," came the response.

I wrote it on the board.

"How many is that?" I asked.

"43," said several.

"Do you think that Paula has 43 stones?" I asked.

"No!!" responded the class.

"Okay, Paula, please come up and tell us how many you have!" I said.

"I have 25 stones," said Paula. She poured her stones into the jar, filling it to the brim, and pressed [25][M+].

"Thank you, Paula. Now how many stones do we have? Shall we look into the calculator's memory, or shall we figure it out without looking?" Several were able to tell me that we had 482. I hit [MRC], and indeed, the number 482 appeared on the display. I wrote 482 stones on the board.

"Thank you for helping me count the stones," I said. "Four hundred eighty-two. That is a lot! It would have taken me a long time to count them without your help."

"Do you know where I got the stones?" I asked.

"No," they said.

"I picked them up on the shore of Lake Superior, which is above the Upper Peninsula of Michigan. Mostly they were underwater when I picked them up. They look much brighter when they are under water. Would you like to see how they look when they are in water?"

"Yes!!" replied the children.

I placed the jar on a child's desk and invited the kids to gather around. I then slowly poured water from a pail onto the stones. "Find a stone and watch it as it becomes wet," I said. The kids were very excited about this. The bright hues of brown, red, orange, burgundy, gray, yellow popped out as the stones were immersed. "That is awesome." "Yes," I said. "It is awesome. If you ever get a chance, take a trip to Lake Superior. It is worth it just to see the stones."

"Thank you again for helping me count. How many did we have in all?" I asked. "482," they responded. I collected the cups and calculators.

An important aspect of this lesson is that each child take his or her own responsibility for counting his or her stones, and that the teacher not 'check' it. As I taught it, I knew an approximate range for the children's stones (12 to 30) and I was able to note that a child's count was within this range. Complete accuracy in counting is not the goal here!

It is a good idea for the teacher to keep her own count of the number of stones currently in the jar, as the children enter their numbers. (This can be done using a separate calculator.) Children invariably mispunch, and the teacher can then correct their mistakes.

CHAPTER 4

Toothpick Numbers

THE LESSON

Decimal notation uses the set of 12 characters: 0 1 2 3 4 5 6 7 8 9 . – (ten digits, a decimal point, and a minus sign). In early grades children see three versions of this set: handwritten, printed, and displayed on four operation calculators and digital clocks. We will call this last version "toothpick numbers."

Reading a calculator's display leads to some systematic but preventable errors. In the first grade, the most common error is confusing 5 and 2, and, less often, confusing 6 and 9. At this stage children reverse numbers and letters, or write them upside down. These errors are rare in the second grade and practically nonexistent later. A more serious and persistent error is ignoring the minus sign or ignoring both the minus sign and the decimal point. (It is very rare for children to ignore just the decimal point but not the minus sign.) This error is due mainly to a lack of understanding of decimal notation. Children who are taught only about whole numbers in very early grades form an incorrect view of decimal notation. They do not consider the decimal point and minus sign as parts of written numbers, but as extraneous or meaningless marks which can be ignored. The decimal point is often treated as a separator between two numbers, and tasks such as, "Add 12.3 and 2.45," are done as follows: 12 + 3 + 2 + 45.

To prevent both types of errors, one should introduce the full decimal notation from the very beginning. This does not mean that children must immediately learn properties of negative numbers or decimal fractions. It is enough if they know that –34.08 is as good a number as 12, with a minus sign and a decimal point being essential parts that cannot be ignored. It is fine if they are told that they will learn more about such numbers in higher grades.

Here is one activity that can be carried out in kindergarten or first grade (or higher). The teacher needs a chart of toothpick digits which must also include a minus sign and a decimal point (see Illustration 4, made on a full sheet of poster board). Children are given toothpicks and calculators, and a stone to represent a decimal point. They enter numbers into their calculators (also using the decimal-point key and the plus-minus key), and copy them with toothpicks and the stone. They may physically move the decimal point (stone) in their toothpick number, and then figure out what operation must be done on the calculator to cause the decimal point to move in the same way. Questions dealing with the structure of digits, such as how many toothpicks are used, should be discussed. (Children are intrigued by finding that 4 uses 4 toothpicks, 5 uses 5, and 6 uses 6.) Having in view a digital clock that displays the same set of digits also helps.

17

REMARKS

Do not ask children to draw toothpick numbers. It could encourage copying digits from the calculator's display instead of writing them using acceptable handwriting. One first grade teacher had her children actually glue down the toothpicks when they formed their numbers.

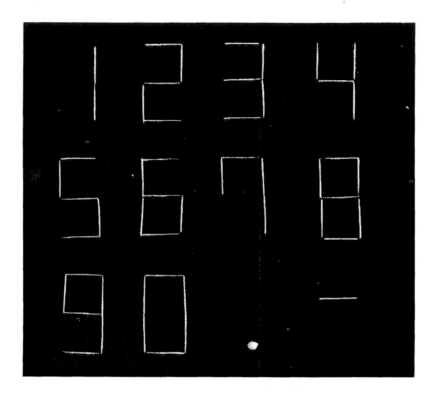

ILLUSTRATION 4. Toothpick numbers.

Follow-up to toothpick numbers

The three basic operations which distinguish between real numbers in decimal notation and whole numbers are:

1. Taking the opposite.
   The opposite of 5.7 is −5.7, the opposite of −5.7 is 5.7, and so on.
   Only 0 is its own opposite.
   Taking the opposite is also described as "changing the sign" of a number.

On the TI-108 the [+/−] computes the opposite.

Remark.
   Notice that this key does not work as a "sign before the digits" key, available on some calculators. In order to enter −5.7 press [5][.][7][+/−] and NOT [+/−][5][.][7].

2. Division by 10.

Division by 10 is seen as moving the decimal point one place to the left, or in the case of integers written without the decimal point and ending with 0, as removing the last 0.

On the TI-108, the combination of keystrokes [/][10] followed by [=][=] ... drives the decimal point to the left.

3. Multiplication by 10.

This is seen as moving the decimal point to the right or appending 0 at the end.

On the TI-108, the combination of keystrokes [M+][10][*][MRC][MRC] followed by [=][=] ... drives the decimal point to the right.

CHAPTER 5

Arithmetic with Manipulatives

THE LESSON

Part I

This lesson has been taught in kindergarten and first grade. In kindergarten, only manipulatives [1], [2], [4], and [8] were used, and they were cut out beforehand by the teacher. She also gave each child a black "mat" (piece of construction paper) on which to make shapes and numbers. Children then knew what to include when they were counting, namely, the manipulatives (square inches) that were on the mat.

1. INTRODUCTION

In the first grade a large amount of time is spent on small whole numbers (less than 30), when children learn basic addition and subtraction facts. This teaching is done mainly with manipulatives, such as decimal blocks, Unifix cubes, counters, Cuisinaire rods, and so on. Here we suggest a different set of manipulatives, which possibly can replace the other ones.

2. MANIPULATIVES

One set of manipulatives consists of five rectangles (three of them are squares) labeled with the numbers 1, 2, 4, 8, and 16. The label describes the area of the rectangle measured in square inches. So the set looks like the ones shown in Illustration 5.

The pieces may be of different colors, and they can be decorated. But the numbers written on them should be legible for the children. Every counting number up to 31 can be uniquely represented as an area (measured in square inches) built from these rectangles. Rectangles are shown below as [1], [2], [4], [8], and [16]:

	1 [1]	2 [2]	3 [2][1]
4 [4]	5 [4][1]	6 [4][2]	7 [4][2][1]
8 [8]	9 [8][1]	10 [8][2]	11 [8][2][1]
12 [8][4]	13 [8][4][1]	14 [8][4][2]	15 [8][4][2][1]
16 [16]	17 [16][1]	18 [16][2]	19 [16][2][1]
20 [16][4]	21 [16][4][1]	22 [16][4][2]	23 [16][4][2][1]
24 [16][8]	25 [16][8][1]	26 [16][8][2]	27 [16][8][2][1]
28 [16][8][4]	29 [16][8][4][1]	30 [16][8][4][2]	31 [16][8][4][2][1]

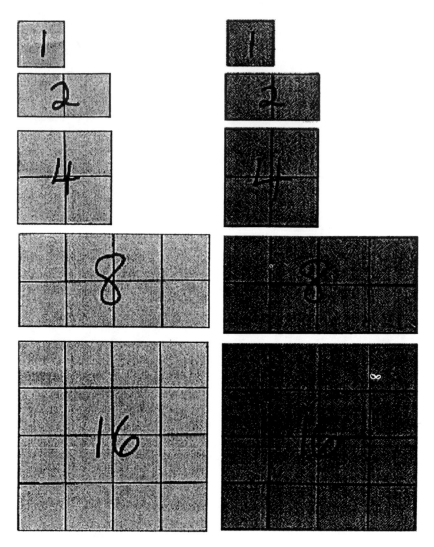

ILLUSTRATION 5. Two sets of five rectangles, each labeled with the numbers 1, 2, 4, 8, and 16.

We recommend that for first grade, each child has two sets (10 pieces), in two different colors, so that up to two numbers up to 31, or one number up to 62, can be represented.

3. CREATING A SET

We suggest that in first grade each child creates his or her own set. The amount of work done by a child should be determined by the teacher. Because reasonable precision is needed, we suggest that the rectangles be precut from stiff colored poster board. Each child gets a large envelope on which he or she writes his or her name, and some rectangles (only a few at one time).

The size of the area of each rectangle is discussed (the smallest is a square inch), and each child writes his or her name on one side of the rectangle and the appropriate number (possibly with help from the teacher) on the other side. The side with the name may be decorated.

This activity should be carried out (possibly for 3 or 4 times) during one week until each child has 2 sets of manipulatives in the envelope.

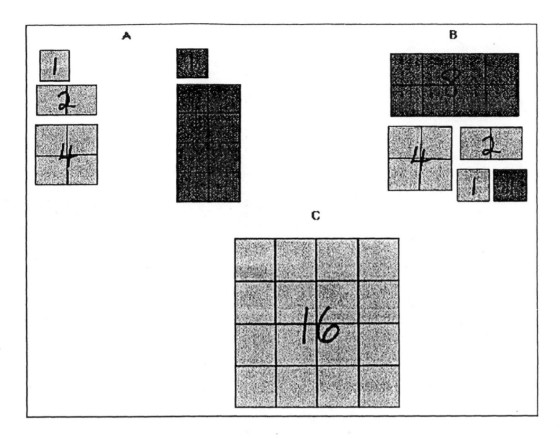

ILLUSTRATION 6. (A) Make 7. Make 9. (B) Add them. (C) Regroup.

4. ACTIVITIES

Any arithmetic activity using any manipulatives may be carried out with this set. We suggest:

- making numbers (e.g., the teacher says, "Make 7. Make 9. Add them. Read the number." (See Illustration 6.)
- counting the total given. (When [4] and [2] are present, the total is 6, not 4 + 2! If needed at first, calculators can be used.)
- asking if the representation of a number is unique
- asking, by using two colors, how many ways I can make a number
- asking what areas I can make: 3 × 3 square inches? 5 × 5? 6 × 6?
- addition (with regrouping)
- subtraction with regrouping
- learning basic addition and subtraction facts
- using the sets to help with mental calculations
- making pictures (a dog, a cat, a human, etc.)

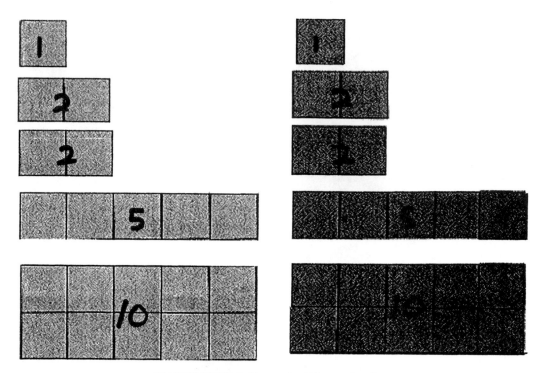

ILLUSTRATION 7. Two sets of five rectangles.

Part II: Another Set of Manipulatives

Above is an example of a simple set of manipulatives which can be used in teaching the concept of regrouping in algorithms for addition and subtraction in base 10. They are rectangles built from squares 1 inch by 1 inch, just as in the materials in Part 1. Each child should have one set of 10 rectangles (see Illustration 7). We suggest (as we did also in Part 1) that children help in making these manipulatives, decorate them, and write the number of square inches on each one (1, 2, 5, 10).

PRELIMINARY ACTIVITIES

(1) Learning about 1 square inch (measuring its sides)
(2) Making different figures and finding their areas by counting or adding
(3) Making different figures of the same area
(4) Making the same figure from different pieces.

MAIN ACTIVITIES

(1) Adding two numbers smaller than 10, with regrouping.

Example: 7 + 9

(a) How to make 2 numbers? [see Illustration 8]
(b) How to add them? [see Illustration 9]
(c) How to regroup? [see Illustration 10]

(6 may also be presented as 5 and 1.)

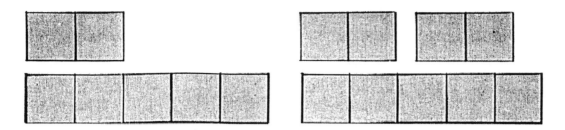

ILLUSTRATION 8. Making 7 and 9.

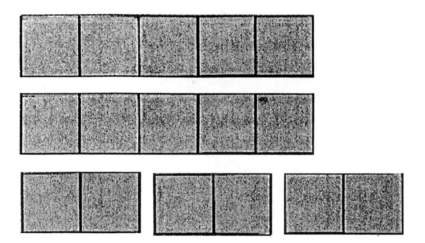

ILLUSTRATION 9. 7 plus 9.

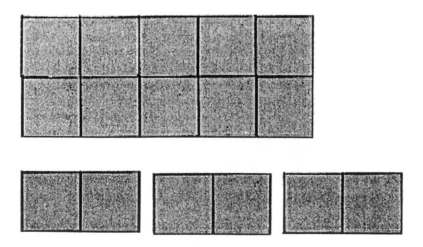

ILLUSTRATION 10. Regrouping.

25

(2) Subtracting a one digit number from 10.

Example:

(a) Regrouping 10

$$10 = 5 + 5 = 5 + 2 + 2 + 1$$

(with manipulatives)

(b) Canceling

$$10 - 4 = (5 + 2 + 2 + 1) - (2 + 2)$$

(with manipulatives)

(3) Subtracting a one digit number from a two digit number. (Regrouping of 10 should now be done mentally.)

COMMENTS

- An explanation of "why" should be done in terms of areas. "We replace one figure by another of the same area."
- The statement that base 10 requires only that we "replace one ten by ten ones" is incorrect and misleading. "You may replace 10 by the sum of any numbers that add up to ten."

CHAPTER 6

Thirty-six Pennies

This lesson has been taught in a second grade class.

PROPS

Each child (and the teacher) should have a cup containing 36 pennies, a magnifying glass, and a calculator.

The teacher should also have an overhead calculator. An extra cup of pennies is also handy, to replace a few which may roll on the floor.

THE LESSON

Part 1. Looking at Pennies

After the pennies, magnifying glasses, and calculators were passed out, I asked the children to take a penny from their cup and look at it. "Who is the person on the penny?" I asked. "You can use your magnifying glass to take a close look if you want." For many, this was the first time they had used a magnifying glass, and they required a bit of help. But soon children were able to focus on the penny using them, and there were several guesses about who the person was ("Bill Clinton," "Ronald Reagan") before they agreed it was Abraham Lincoln. We also read the words on the front of the coin, "Liberty" and "In God we trust."

"Turn the penny over for a minute. Do you know what that building is?" I asked.

"The White House!" said several.

"No," I said, "but you are in the right city!" No one knew that it was the Lincoln Memorial. "Take your magnifying glass, and focus on the center of the building, between the middle two columns," I said. "What do you see? You may need to get a new shiny penny from your cup, in order to be able to see it."

"Wow, I do see something!" said Sally. "It looks like someone sitting in a chair."

"Who do you think it is?" I asked.

"Maybe it is Abraham Lincoln," said Arnetha.

"Right," I said. "Has anyone in this class been to the Lincoln Memorial and seen the statue of Mr. Lincoln sitting in a chair?" No one raised his or her hand. "Well, it's worth the trip," I said. "And can you read the words on the back of the penny?" "One Cent" and "United States of America" were easy. But no one tackled "E pluribus unum." "It is not English," I said. "It is Latin. And it means 'one among many.' You will learn more about it in history, but the United States is one country formed with people from many backgrounds, so it is a fitting thing to put on our penny."

"Now turn your penny back over, to the head of Mr. Lincoln. Do you see the number below his chin?" I asked.

"Yes!"

"Can you read it?"

The children read aloud, "1980!" "1994!"

"What does the number mean?" I asked. "It was when Abraham Lincoln was born!" said Jack.

"No, it was when he died," said Arnetha.

"Wait a minute," I said. "We have so many dates on all the pennies. How can they all be about Mr. Lincoln?"

"They are not about him," said Aric. "They tell when the penny was made."

"Right," I said. "Each penny has a date on it, and the date is the year the penny was made. So now I have a question for you. Who has the oldest penny in this class? Gently dump out your pennies—don't let them roll on the floor—and take a look at the dates. If you think you have an old penny, tell me its date, and I will write it on the board."

This activity caused quite a stir. The children called out dates (including 1996 and 1995). Carlos said, "A 1996 penny is not very old. It was just made last year."

"Oh, I get it," said Sally. "We want 'old' pennies, not new ones!"

"That's right," I said. "Smaller numbers mean older pennies."

One child found a 1960 penny, and another (Elizabeth) found a 1948 penny. I recorded about 10 dates on the board and asked the children which one was on the oldest penny. They selected 1948. The 1948 penny is a "wheat" penny, it turns out. There is no Lincoln Memorial on its back side, but sheaves of wheat instead. I pointed this out to Elizabeth, who proudly showed her penny to several classmates.

"I have a question for you," I said. "How old is Elizabeth's penny?"

"Very old!" said several.

"Well, how can we figure out how many years old it is? What year is this?"

"1997," they said.

"And what year was Elizabeth's penny made?"

"1948."

I wrote on the board:

    1997
    −1948
    ‾‾‾‾‾

"Why don't you try that on your calculator?" I asked. I wrote on the board:

[1997][−][1948][=]

"49!" said Elizabeth, together with several others.

"49 what?" I asked. "How old is the penny?"

"49 years old! Wow, that is old," said Elizabeth.

Part 2: Counting Pennies by Making Arrays

I gently dumped out my pennies onto the overhead. "Now let's have some fun with the pennies," I said. "First we need to know how many pennies we have. I'm going to line my pennies up in rows of 10, and see how many I have." I lined mine up as shown in Illustration 11.

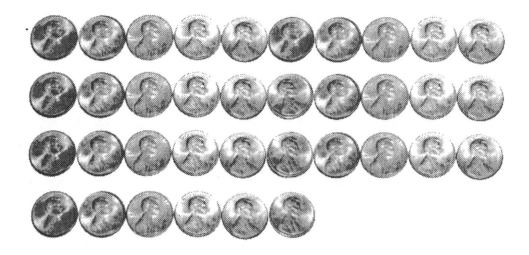

ILLUSTRATION 11. Thirty-six pennies.

"Can you line yours up like this?" I watched to see that the children were able to do it. "How many do we have?" I asked.

"10, 20, 30," counted Jason. "And some left over."

"How many are left over?" I asked.

"Six," he said.

"So how many altogether? If you need help, you can use your calculator."

"I think there are 36," said Carla.

"How many agree?" Quite a few hands went up. "Try [30][+][6][=] on your calculator," I said, and I wrote it on the board.

"Yes, we have 36," said several.

"I want to show you something cool," I said. "Do you know about the 'times' key?" Some did, but most did not. "Well, it can be very helpful in a situation like this," I said. "Let's look at our pennies. How many full rows of 10 do we have?"

"Three."

I wrote [3][*][10] on the board. "And how many extra?"

"Six."

I wrote on the board, [+][6][=]. "We have 3 rows of ten, plus six. So we need to press [3][*][10][+][6][=]. Try it! What do you get?"

"36!" they said.

"Well, that is pretty interesting," I said. "I am going to rearrange my pennies just a little bit. Watch me, and then you can try it too. Then we will see if we still have 36 pennies. I am going to take one penny from the end of each of my three full rows, like this, and move them down to my bottom row." My picture now looked like this:

○○○○○○○○
○○○○○○○○
○○○○○○○○
○○○○○○○○

"Can you do it?"

"Yes!"

"How many full rows do you have now?"

"Four."

"And how many in each row?"

"Nine."

"Try [4][*][9][=] on your calculator!" (I wrote it on the board.)

"36!" The children thought this was awesome.

"Okay, now let's continue. I'm going to take one penny from the end of each of my full rows, and put it down below, on a new row. My new picture looks like this:

○○○○○○○
○○○○○○○
○○○○○○○
○○○○○○○
○○○○

"Try it. How many full rows do I have?"

"Four!"

"And how many in each full row?"

"Eight!"

"And how many left over?"

"Four!"

"So what keys do I need to press? Am I going to get 36 again?" I wrote on the board, [4][*][8][+][4][=]. "Try it!"

"Thirty-six!" said the children.

"What do you think I'm going to do next?" By this time the children knew. I would take one penny off the end of each full row, and put it at the bottom. "In case you want to know what mathematicians might call the pictures we are making, they are 'arrays,' " I said. I wrote *array* on the board. My next array looked like this:

○○○○○○
○○○○○○
○○○○○○
○○○○○○
○○○○○○
○

"And mathematicians call a sequence of keystrokes a 'program,' " I said. (I wrote *program* on the board.) "What is the program for this array?" Now several children could say the keystrokes: [5][*][7][+][1][=].

"Let's just continue. Now I'll take a penny from each of my full rows. How will it look?"

○○○○○○
○○○○○○
○○○○○○
○○○○○○
○○○○○○
○○○○○○

The program: [6][*][6][=].

We continued, with the following programs (and each time we changed our array):

[7][*][5][+][1][=]
[9][*][4][=]
[12][*][3][=]
[18][*][2][=]

And finally (!):

[36][*][1][=].

"Boy, there are a lot of ways to count to 36!" said several.

"Well, before we quit, I want to show you one more way" I said. "This time, I will make four rows, and in each row there will be three stacks of three pennies each, like this" [(○○○) means a stack of 3 pennies]:

(○○○)(○○○)(○○○)
(○○○)(○○○)(○○○)
(○○○)(○○○)(○○○)
(○○○)(○○○)(○○○)

"Now, what do you think the program is? We have 4 rows of 3 stacks of 3 pennies." The children suggested [4][*][3] but then became stuck. "We need to times it once more, but by what?" I asked. "How many pennies are in each stack?"

"Three," they said.

"Let's try 3, then," I said:

[3]*][4][*][3][=]

"Thirty-six!" they said in unison. "That is really cool!"

"That's enough for today," I said. "I bet you never knew there are so many ways to count to 36!"

"Never!" said the children.

We collected the cups of pennies.

CHAPTER 7

Subtraction Table

This lesson has been taught in a second grade class.

THE LESSON

Task. Construct a subtraction table for the numbers 1, 2, 3, 4, and 5.
Tools and props: Paper with a 1 in. by 1 in. grid of dots, rulers, crayons.

(1) Prepare the table (see Illustration 12).
(2) Fill out the table. Use a calculator to subtract a top number from a number on the left (for example, [3][−][4][=]; the display shows −1), and write the result in the appropriate square (see Illustration 13). Do not use automatic mode (the constant-key feature).
(3) Use crayons to decorate the table.

When we tried this lesson, children were quite delighted with the patterns they discovered ("Did you see the whole row of zeroes?" "There are fours in two corners, but one has a minus sign!")

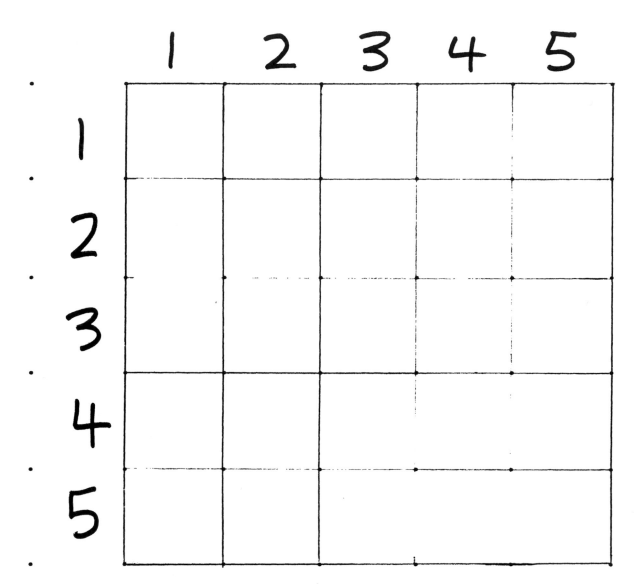

ILLUSTRATION 12. Preparing a subtraction table.

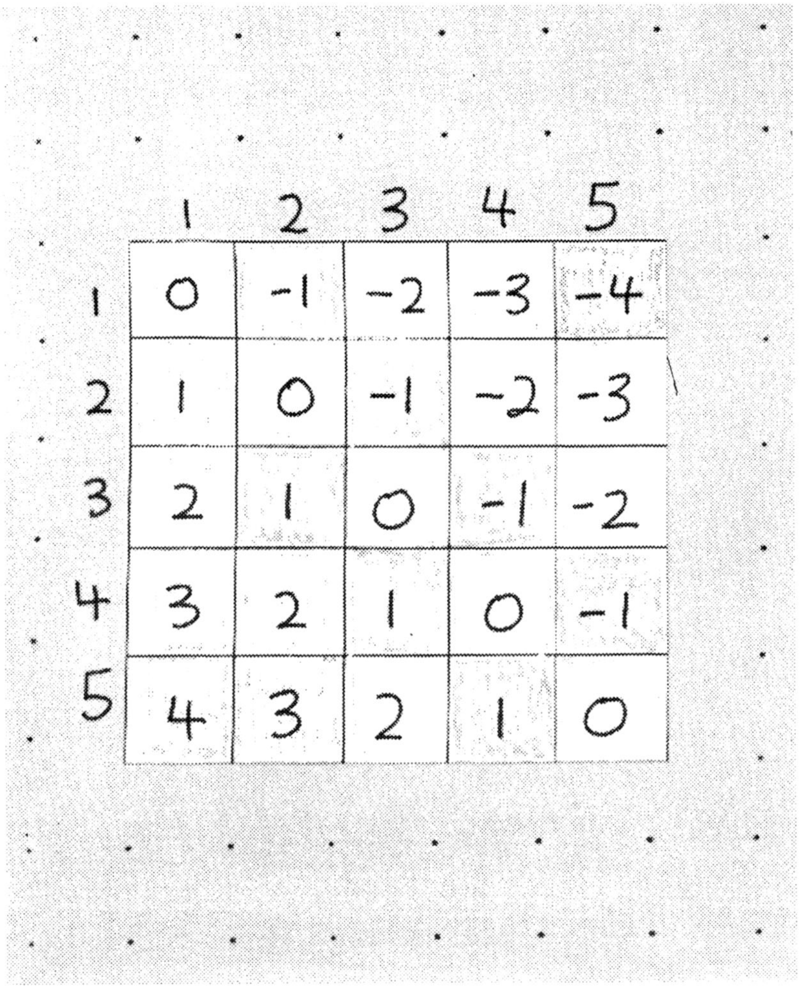

	1	2	3	4	5
1	0	-1	-2	-3	-4
2	1	0	-1	-2	-3
3	2	1	0	-1	-2
4	3	2	1	0	-1
5	4	3	2	1	0

ILLUSTRATION 13. Subtraction table.

CHAPTER 8

Squirrels' Race

PROPS

A handout with the Squirrels' Race story on it, together with an overhead transparency of the story (see end of lesson). Zip-lock bags of about 25 acorns (or hazelnuts) and seven small paper plates for each group of children. Calculators are optional (we did not use them). Groups can consist of two to seven children.

THE LESSON

Today I saw an expert teacher in action. She was Joyce Watson, second grade teacher at Jornada Elementary School, Las Cruces, New Mexico. She first assigned each child to a group; there were five groups, each with four to six children. She placed the transparency on the overhead and asked the children to read aloud with her:

Squirrels and Acorns

There were 7 squirrels who decided to have a race.
They agreed that each one would get something as a prize.
And for squirrels the best prizes are acorns.
And the squirrels decided that the faster ones should get more acorns than the slower ones. (Each squirrel should get a different number of acorns.)
How many acorns did they need for prizes?
Can you help them?

"Okay," she said, "Can we help them? What are we supposed to do?"
"Decide on the prizes for the squirrels," said Dominique.
"And how are the prizes to be distributed?" she asked.
"The one who wins gets the most acorns," said Hillary.
"What about the others?" "Faster ones should get more than slower ones," said Jason.
"Do you think we can figure out how to distribute the acorns?" she asked. "Do any two squirrels get the same number of acorns?"
"No," said Brian. "Each squirrel gets a different number."
"Okay, I have some things to pass out to you that might help. Each group gets a copy of the story, just so you can refer to it, seven plates, one for each squirrel, and a bag of acorns. If you need more acorns, I have plenty, so just ask me for them. Maybe you can

just put acorns on each of the plates. When you think you have figured out how many acorns each squirrel should get, raise your hand and we will discuss it."

She passed out the nuts and plates, and the classroom started to buzz with children's voices: "Let's see, shall we dump our acorns out?" "How many acorns do we have?" "What shall we do with the plates?" "Maybe we should line them up."

"Yes," Mrs. Watson said, "Each plate can represent a squirrel."

The children continued to ponder: "I think we might need to write something down, or maybe draw a picture." "Hmm, this is hard. I don't know how to start." "Maybe we can just put some acorns on a plate, and start from there." "I know that the squirrel who comes in first gets the most." "And the squirrel who comes in last gets the least!" "But how to do it?"

The light began to dawn on several children. "We can't put the same number on two plates. So how can we arrange the nuts?" "Hmm, maybe one here, two there. . . ." "But we don't have enough. Let's ask for more." Children began to request one more nut or two more; one group asked for 28 more. Mrs. Watson obligingly filled all requests. "It might help," she said, "if you arranged your plates in a straight line."

In one group I overheard this conversation: "Okay, let's put one acorn on every plate. Now take one plate away, and put one more acorn on each plate that is left. Now take another plate away, and put another acorn on each plate that is left. If we keep doing this, maybe we will solve the problem."

"Remember, when you think you know how to distribute the acorns, raise your hand and I will write your solution on the board," said Mrs. Watson. "Also, figure out how many acorns you are using altogether."

One by one, the groups began to raise their hands. When all five groups were satisfied that they had a solution they wanted to present, Mrs. Watson called on a member of the first group.

"Casey, can you tell the class how your group arranged the nuts?" She wrote on the board as he spoke:

last place (7th)	1 nut
next-to-last place (6th)	2 nuts
5th place	3 nuts
4th place	4 nuts
3rd place	5 nuts
2nd place	6 nuts
1st place	7 nuts

She actually recorded Casey's solution from bottom to top, so when it was finished, it looked like:

1st	7
2nd	6
3rd	5
4th	4
5th	3
6th	2
7th	1

Casey used his group's layout of plates and nuts to guide him as he spoke. It was difficult for him to count backwards (saying the places) and forwards (saying the number of nuts) at the same time. At "4th place, 4 nuts" he said, "I forgot in which direction I was going!" "Yes, it gets sort of confusing, doesn't it?" said Mrs. Watson.

"Does it look like Casey's group has a solution?" she asked.

"Yes," said the children.

"How many nuts does his group need for prizes?" she asked.

"29," said some.

"28," said others.

"Well, how can we know for sure?"

"Add them up," they replied.

"Okay, 1 plus 2 is what?"

"3."

"3 + 3 is . . ."

"6."

"6 + 4 is . . ."

"10."

"10 + 5 is . . ."

"15."

"15 + 6 is . . ."

"21."

"21 + 7 is . . ."

"29!"

"No, 28!"

"How can we know for sure?"

"For sure, it is 28," they decided. "21 + 7 is 28."

"I will show you another way to count them," she said. As she spoke, she made V-shaped brackets on the board that grouped together the numbers she was referring to. "Look, 1 + 7 is what?"

"8."

"2 plus 6 is what?"

"8."

"3 plus 5 is what?"

"8!"

"What is 3 times 8?"

"24!"

"Plus how many more, left over here in the middle?"

"4."

"24 plus 4 is?"

"28."

The children were very good at mental math. They did not use calculators!

"Did any other group get this solution?" she asked. Children from two other groups raised their hands. "Good job," she said. "Did anyone get a different solution? Cindy, what did your group come up with?" She wrote on the board as Cindy spoke.

1st	8
2nd	7
3rd	6
4th	5
5th	4
6th	3
7th	2

"Does this solution work?" she asked.

"Yes," replied the children.

"How many nuts does Cindy's group need?" she asked.

"Easy! 35."

"Why?"

"You just add one extra nut for each squirrel, and there are seven squirrels, and 7 plus 28 equals 35," said Jason.

"Wow," said Mrs. Watson. "That is impressive. I can show you a nifty way to add them up, to check." As before, she grouped 2 + 8, 3 + 7, and 4 + 6. "How much is that?" she asked.

"30."

"And how much in the middle?"

"5."

"So how many nuts altogether?"

"35!"

There was still another solution from the final group. "We did it like this," said Veronica. "We gave nine to the first squirrel." As she spoke, Mrs. Watson again wrote on the board:

1st	9
2nd	8
3rd	7
4th	6
5th	5
6th	4
7th	3

"What do you think? Does this solution work?"

"Yes."

"Well, how many nuts does Veronica's group need?"

"45," responded several.

Mrs. Watson wrote "45," and said, "Well, let's check. How much is 3 + 9?"

"11," said several.

"12," said others.

"Jennifer, will you please put up 9 fingers?" Jennifer stuck up her fingers as requested. "And here are three of my fingers," said Mrs. Watson. "How many fingers are there? Let's count."

The children chimed in, "Nine, ten, eleven, twelve."

"So how much is 3 + 9?"

"12."

"4 + 8 = ?"

"12."

"5 + 7 = ?"

"12."

"So we have 3 times twelve. How much is that?" She wrote on the board,

$$
\begin{array}{r}
12 \\
12 \\
+12 \\
\hline
\end{array}
$$

"36," responded the children.

"And we have 6 more left in the middle."

"That's 42," they said.

"Well, that's pretty close to 45," she said, as she corrected the total to make it read 42.

"Let's see, we have three great solutions here. I guess nuts were more scarce in the forest where the squirrels used only 28, and they were very plentiful in the forest where they needed 42.

"But just a minute, the story isn't over yet—I forgot to tell you something. When the squirrels went out to hunt for nuts for prizes, they actually found 31 nuts. And they wanted to use all the 31 nuts for prizes, not 28 or 35 or 42. Can you tell them how they should distribute the 31 nuts?"

"Oh, no," said several. "This is going to be hard!"

"Well," said Mrs. Watson, "Maybe you should begin with exactly 31 nuts, and see if you can lay them out on the plates. I'll come around and give you more nuts if you need them, or I'll take back nuts that you don't need. Remember, the same rules apply: no two squirrels get the same number of nuts."

Some groups had no idea how to begin. When they got their 31 nuts, they dumped them altogether on one plate. One group kept their layout for 28 nuts and tried to think how to change it. Several solutions were offered, but when Mrs. Watson came around to check, she found that two (or more) plates contained the same number of nuts. "That's not allowed," she said. "How can you change it to make it fit the rules?"

The first solution offered by a group was:

1st	8
2nd	7
3rd	6
4th	4
5th	3
6th	2
7th	1

The class checked to see that the total was 31.

Next a group offered:

1st	9
2nd	7
3rd	6
4th	4
5th	3
6th	2
7th	1

But one child pointed out, "That can't work, because it is one bigger than 31; the 8 has been changed to a 9."

"Right," said Mrs. Watson. "How can we change it to make it into 31?"

"Easy," said Carla. "Just change the 6 to a 5."

Mrs. Watson did so; the solution read:

1st	9
2nd	7
3rd	5
4th	4
5th	3
6th	2
7th	1

"Does that work as a solution? Let's check: $1 + 2 + 3 + 4 + 5 + 7 + 9 =$ what?"

"31!"

ILLUSTRATION 14. How many acorns are needed for seven squirrels?

"We have another solution!" said Charles. "We gave 10 to the first place squirrel. Then we gave 6, 5, 4, 3, 2, and 1." Mrs. Watson wrote:

1st	10
2nd	6
3rd	5
4th	4
5th	3
6th	2
7th	1

"Does the solution of Charles's group total 31?" asked Mrs. Watson.

"Yes, because 10 + 6 is the same as 9 + 7, and we don't even need to check the rest," said Brenda.

"Great," said Mrs. Watson. "Well, we have three different solutions here that use 31 nuts. Let's look at them."

1st	8	9	10
2nd	7	7	6
3rd	6	5	5
4th	4	4	4
5th	3	3	3
6th	2	2	2
7th	1	1	1
	31	31	31

"Suppose you were in charge of the race and had to decide which way to distribute the prizes. Which would you choose?" The children had definite opinions about this.

"The first way is the most fair," said Cindy.

"Yes, but the last way gives the first place winner the most, " said Carla.

"Let's vote," said Mrs. Watson. The vote was 2 to 2 to 21. The children preferred to reward the first place winner with the most acorns. "Why did you vote that way?" Mrs. Watson asked.

"The first place winner should get the biggest reward," they said.

"Well, if you were in the race, and you came in second, would you still want that distribution?" she asked. Most replied yes, because it would make them work harder to try to come in first in the next race. Others were not so sure.

"I am glad you could help out the squirrels," said Mrs. Watson. Will you put the nuts back in the bag?" She then collected the plates and nuts.

COMMENT

Having seven plates to lay out, to try out different groupings for the nuts, is essential in this lesson (see Illustration 14).

Squirrels and Acorns

There were 7 squirrels who decided to have a race.

They agreed that each one would get something as a prize.

And for squirrels the best prizes are acorns.

And the squirrels decided that the faster ones should get

more acorns than the slower ones.

(Each squirrel should get a different number of acorns.)

How many acorns did they need for prizes?

Can you help them?

CHAPTER 9

A Modern Abacus

INTRODUCTION

The goal of this lesson is to create a computational toy, a small abacus on which one can calculate with negative numbers and decimal fractions. This toy helps to get some insight into positional notation and the role of negative numbers in arithmetic. The abacus at the end of the unit has been used in upper elementary and middle school classrooms. We cut square tokens from black and red poster board.

THE LESSON

Decimal notation is based on the sequence

$$\ldots 1000 \quad 100 \quad 10 \quad 1 \quad 1/10 \quad 1/100 \ldots$$

The numbers are formed by multiplying the elements of this sequence by the digits

$$0, 1, 2, 3, 4, 5, 6, 7, 8, 9$$

and adding the results.

Example 1

$$2*100 + 0*10 + 7*1 + 7*(1/10) + 9*(1/100)$$

is written as 207.79, with the point (.) indicating the position of 1 in the original sequence.
Binary notation is based on the sequence
$$\ldots 8 \quad 4 \quad 2 \quad 1 \quad 1/2 \quad 1/4 \ldots$$

Only two digits are needed, 0 and 1.

Example 2

$$1*8 + 0*4 + 1*2 + 0*1 + 1*(1/2) + 1*(1/4)$$

is written in base 2 as 1010.11, where again (.) indicates the location of 1 in the sequence.
In both cases, only positive numbers are constructed, so negative numbers are indicated by putting a − sign in front.

Example 3

$$-23.5 = (-1)*(2*10 + 3*1 + 5*(1/10)) = (-2)*10 + (-3)*1 + (-5)*(1/10)$$

In 1726, J. Colson observed that one can use different sets of digits in decimal notation. For example, the following set of 10 digits,

−4, −3, −2, −1, 0, 1, 2, 3, 4, 5,

allows you to write all positive and negative numbers without an extra − sign in front. He also noticed that if we use this system, arithmetic operations become slightly easier. (You need the multiplication table only up to 5; there are fewer carries in addition; you do not need a separate algorithm for subtraction.)

You also may use larger sets of digits. For example,

−9, −8, −7, −6 ,−5, −4, −3, −2, −1, 0, 1, 2, 3, 4, 5, 6, 7, 8, 9.

Each number can now be written in many different ways, which may sometimes be convenient, but which makes a comparison of numbers more difficult.

Decimal notation is called positional because it uses the same digits for units, tens, tenths, hundreds, . . ., and their actual values are determined by their position in a numeral. Positional notations are related to calculation with the Roman abacus, where the value of a token depends solely on its position on the counting board.

New Notation and a Toy Abacus

Consider the following sequence:

. . . 100   50   20   10   5   2   1   1/2   1/5   1/10   1/20   1/50   1/100 . . .

and three digits, −1, 0, 1.

EXAMPLE

$$1*50 + -1*20 + 0*5 + 1*1 + -1*(1/2)$$

would be written

1−101.−1,

and it would be equal to $50 - 20 + 1 - .5 = 30.5$ in the usual decimal notation.

Conversely, every positive or negative decimal can be written in this new notation. Try to figure out how! This is rather tricky, so do not give up easily. Many numbers can be written in many different ways.

The Toy Abacus (see the board at the end of the unit)

You should use two kinds of tokens: a black token for 1, and a red token for −1. Lack of a token indicates 0.

EXAMPLE 1

To represent the number 30.5 from above, put black tokens on 50 and .5, and put a red token on 20. (There are other ways to represent 30.5. For example, put black tokens on 50 and 1, and put red tokens on 20 and .5 .)

EXAMPLE 2

Red tokens on 20 and .2 and black tokens on 2 and .1 give
$$-20 + 2 - .2 + .1 = -18.1.$$

(There are other ways to get -18.1. Can you find one?)

How to add. Put both numbers together and "regroup." When you put numbers together, you may pile several tokens into one place. Some examples of regrouping: Red and black tokens on the same place cancel. Two tokens of the same color on 5 can be exchanged for one token on 10. Two black tokens on 2 can be exchanged for one black on 5 and one red on 1.

How to subtract. Add the opposite number. To form the opposite, replace red tokens by black ones, and black ones by red.

You may even multiply and divide numbers on your toy abaci. But you will have to use several boards (one for each factor, one for the result, and one extra to help with the mental part of the computation.)

.05	.5	5	50	500
.02	.2	2	20	200
.01	.1	1	10	100

CHAPTER 10

Modern Counting Boards

Below are some challenging activities that provide training in mental addition and subtraction. They are suitable for upper elementary and middle school.

THE LESSON

Five boards are given in Illustrations 15–19. Each board consists of nine squares, labeled with numbers as shown. A good size for each square is 2.5 by 2.5 inches. A blank board is given at the end of this lesson. It may be photocopied, and children can write their own numbers.

To represent a number on a board, you need tokens with two sides having different colors, say yellow, Y, and red, R. Y has a value of 1 and R a value of −1. If you put a token on a square, you multiply its value by the value of the square. The total of these "positional" values of tokens gives you a number on the board.

Example 1

Using Illustration 20, we have the number 432 on this board: 1*500, −1*100 in the first column, 1*80, −1*50 in the second, 1*1, 1*1 in the third.

Total 500 − 100 + 80 − 50 + 1 + 1 = 432.

To do this task, children do not have to know either negative numbers or multiplication. It is enough that they know that yellow means that you have to add the number, and red means that the number should be subtracted.

We try to represent the whole numbers from 0 to 999 in such a way that:

(1) Each column represents one digit of the number.
   Getting 55 by putting Y on 50, Y on 10, and R on 5 is "illegal", in spite of the fact that 50 + 10 − 5 = 55.

(2) There should be at most 2 tokens in each column.
   Y on 8 and R on 5 represents 3 "legally", but three Y's on 1 are "illegal", in spite of the fact that they give the correct value.

Each whole number from 0 to 999 can be represented on each of the five boards legally. (There is no other board with this property.) Thus you need at most 6 two-sided tokens to represent a number.

The board in Illustration 15 is the easiest. Those in Illustrations 16 and 17 are more difficult. The two boards in Illustrations 18 and 19 are the most difficult.

700	70	7
200	20	2
100	10	1

ILLUSTRATION 15.

800	80	8
300	30	3
100	10	1

ILLUSTRATION 16.

800	80	8
500	50	5
100	10	1

ILLUSTRATION 17.

500	50	5
400	40	4
300	30	3

ILLUSTRATION 18.

500	50	5
400	40	4
200	20	2

ILLUSTRATION 19.

800	80 Y	8
500 Y	50 R	5
100 R	10	1 Y Y

ILLUSTRATION 20. 432 is on the board.

Activities

Children should work with one board at a time, until the activities become quite automatic (and not challenging). Because this is skill training in mental computations, sessions that are frequent, but not very long, are probably the best. Paper and pencil may be allowed only for writing the results, or not at all. (The results may be shown or be presented orally.) Calculators are not needed, but they may be used for limited purposes at the teacher's discretion.

For all the activities listed below, each child needs one board and 8 tokens. At the very beginning children may work in groups to help each other. But remember that an individual and not a group acquires a skill. The teacher has a transparency of a board to be shown on the overhead projector, and transparent tokens of two colors. (The teacher's colors should be the same as the colors on the two sides of the children's tokens.)

ACTIVITY 1

Putting a number on a board, and reading a number:

(1) The teacher shows legal configurations on the projector. Children copy them on their board and read and possibly write them in decimal notation.
(2) The teacher names whole numbers (up to 3 digits) or writes them on the blackboard. The children figure out how to represent them legally.

Please note that
- Sometimes there are two different legal representation of a digit on a board.
- The children's work must be checked so that any systematic errors are caught and explained.

ACTIVITY 2

Regrouping tokens to make a number legal:

(1) The teacher shows illegal configurations on the overhead projector. The children find their values. (This is a place where the teacher may allow children to use calculators.)
(2) The teacher shows illegal configurations on the projector. The children rearrange the tokens to get a legal representation of the same number. (No calculators! The work should be done mentally.)

ACTIVITY 3

Adding and subtracting three-digit numbers on a board:

(1) The teacher puts an addition or subtraction problem on the blackboard.
(2) The children represent the first number legally on their board, and then add (or subtract) the second number, one digit at a time, rearranging the tokens to make the representation legal. (Any order of adding, right to left, or left to right, is fine.) They may be required to write down the answer in decimal notation, to be checked later by the teacher.

Example 2

$$342 + 269$$

Put 342 on the board, $300 = 800 - 500$, $40 = 50 - 10$, $2 = 1 + 1$ (see Illustration 21); add $200 = 100 + 100$ (Illustration 22); regroup, $800 - 500 + 100 + 100 = 500$ (Illustration 23); add $60 = 50 + 10$ (Illustration 24); regroup, $10 - 10 + 50 + 50 = 100$ (Illustration 25); add $9 = 8 + 1$ (Illustration 26); regroup, $8 + 1 + 1 + 1 = 11 = 10 + 1$ (Illustration 27). The answer is: 611.

800 Y	80	8
500 R	50 Y	5
100	10 R	1 Y Y

ILLUSTRATION 21. 342 is on the board.

800 Y	80	8
500 R	50 Y	5
100 Y Y	10 R	1 Y Y

ILLUSTRATION 22. Add 200 to 342.

800	**80**	**8**
500 Y	**50** Y	**5**
100	**10** R	**1** YY

ILLUSTRATION 23. Regroup: 800–500 + 100 + 100 = 500.

800	**80**	**8**
500 Y	**50** Y Y	**5**
100	**10** RY	**1** YY

ILLUSTRATION 24. Add 60 = 50 + 10.

800	80	8
500 Y	50	5
100 Y	10	1 Y Y

ILLUSTRATION 25. Regroup: 10 − 10 + 50 + 50 = 100.

800	80	8 Y
500 Y	50	5
100 Y	10	1 Y Y Y

ILLUSTRATION 26. Add 9 = 8 + 1.

800	**80**	**8**
500 **Y**	**50**	**5**
100 **Y**	**10** **Y**	**1** **Y**

ILLUSTRATION 27. Regroup: 8 + 1 + 1 + 1 = 11 = 10 + 1. The answer is 611.

REMARKS

- All the regrouping should be done mentally.
- This is drill, but it provides a considerable intellectual challenge.
- Children should not be rushed, but a rather brisk pace should be maintained. Those children who have problems should be helped individually later.
- Again, the work should always be checked for errors, either by the teacher or by other children.

CHAPTER 11

Abax: A Computing Device

We describe here a "modern counting table" which we call by an ancient Greek name, Abax. In Part I we show how it can be used for addition and subtraction, and as a teaching tool for explaining base ten notation, including decimal fractions. In Part II we show how it was used in a fourth grade class, and in Part III we show a game that can be played with it.

THE LESSON

Part I

A number is represented by tokens put on a counting board. Examples of two boards used are given in Illustrations 28 and 29. The tokens represent powers of 10. We use three kinds of tokens.

> [1], [10], [100] – for basic arithmetic in the range 0 to 999.
>
> [1¢], [10¢], [$1], [$10] – for lessons using money.
>
> [.001], [.01], [.1], [1], [10], [100], [1000] – for advanced lessons using decimal fractions.

The tokens must be easy to move, and all tokens in each set must fit into one rectangle on the board. We used wooden tokens, three star-shaped pieces of different sizes for the first set, medium sized round plugs for the second, and small round plugs for the third set. Each plug in a set was painted with a different color, and its value was clearly written on it.

Examples of representations of numbers are given in Illustrations 30–32.

GENERAL DESCRIPTION

We look at a power of 10 (a unit of some size) and a number between 0 and 9 (a multiplicative factor). We represent it on the board by putting a token (unit) on the rectangle (factor). Any number is the sum of such products.

Addition and subtraction (we will show it only for a basic set): The only skill needed for addition and subtraction is counting up to 10.

EXAMPLE 1

Add 276 to 307. (See Illustration 30.) Begin with 307. Walk the [100] token 2 places clockwise to get the configuration shown in Illustration 33.

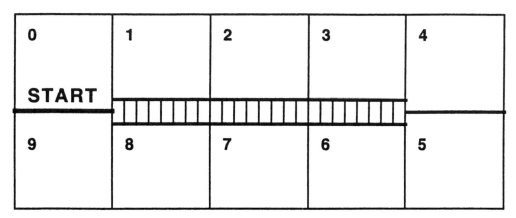

ILLUSTRATION 28. One possible board design for an Abax.

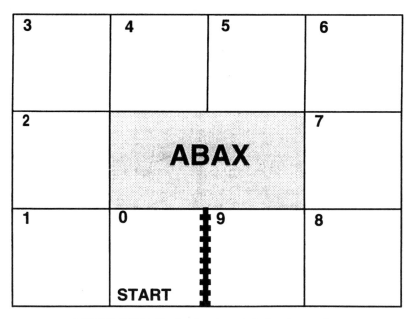

ILLUSTRATION 29. Another board design for an Abax.

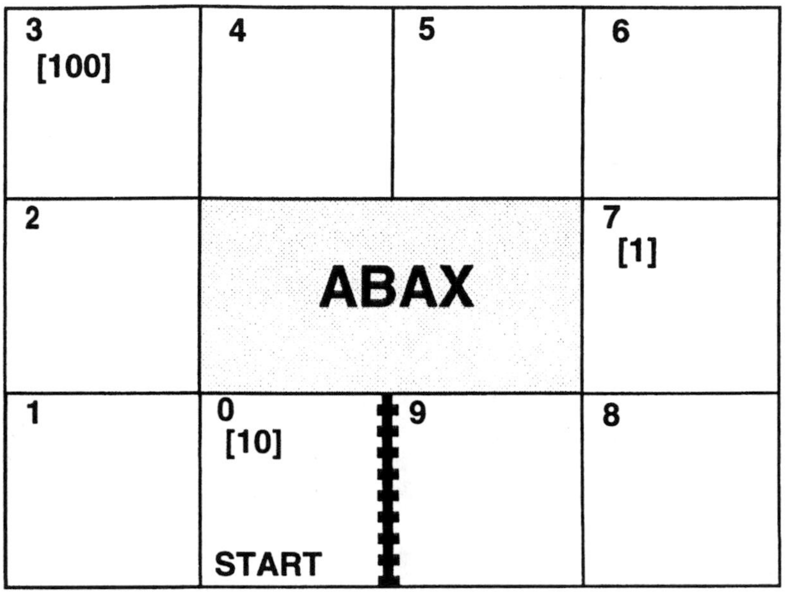

3 [100]	4	5	6
2	**ABAX**		7 [1]
1	0 [10] ⋮ 9		8
	START		

ILLUSTRATION 30. The number 307 is represented.

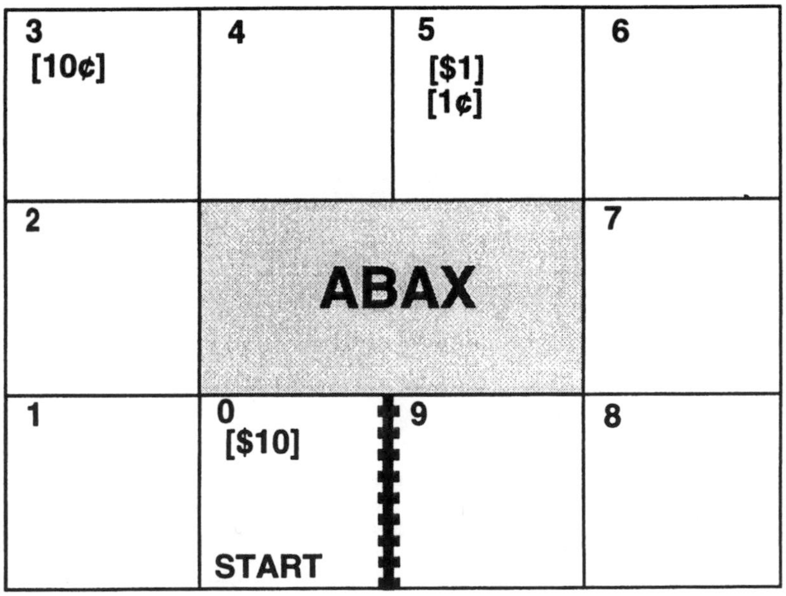

3 [10¢]	4	5 [$1] [1¢]	6
2	**ABAX**		7
1	0 [$10] ⋮ 9		8
	START		

ILLUSTRATION 31. $5.35 is represented.

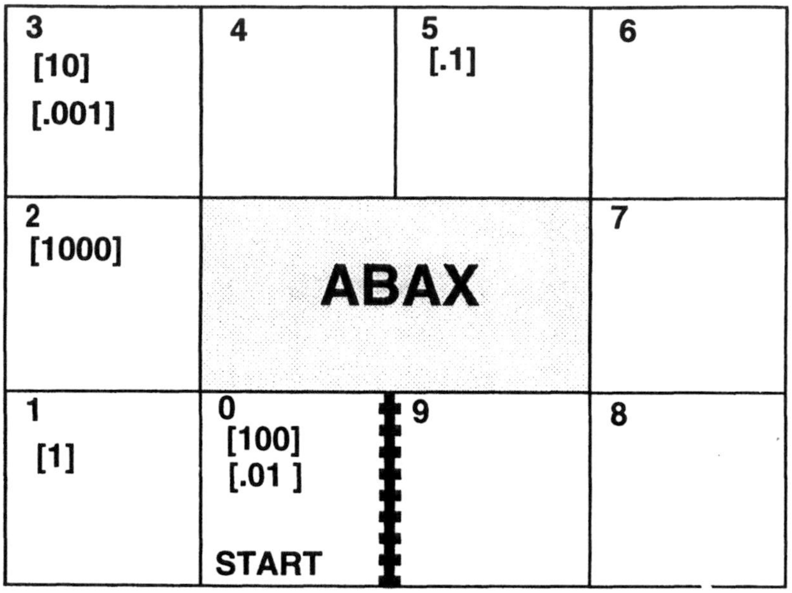

3 [10] [.001]	4	5 [.1]	6
2 [1000]	ABAX		7
1 [1]	0 [100] [.01] START	9	8

ILLUSTRATION 32. 2031.503 is represented.

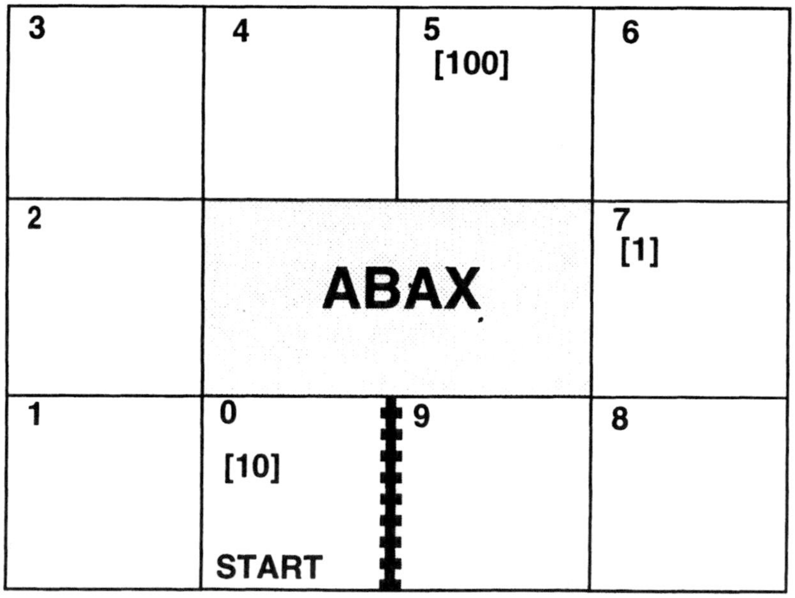

3	4	5 [100]	6
2	ABAX		7 [1]
1	0 [10] START	9	8

ILLUSTRATION 33. Begin with 307 (shown in Illustration 30). Walk the [100] token 2 places clockwise to get 507, shown above.

Walk the [10] token 7 places clockwise to get the configuration in Illustration 34. Walk the [1] token 7 places clockwise to get the configuration in Illustration 35, and because your [1] token crossed the START line (marked ▐▬▬▬▬▬▬▌), advance your NEXT HIGHER token ([10]) one place (Illustration 36).

Subtraction is achieved by walking tokens counter-clockwise, according to the same rules.

Comments

Suggested uses of an ABAX.

- It is really a very powerful arithmetic device so it can be used in early grades (K–2) to solve more complex problems involving addition and subtraction of whole numbers and money.
- It can be used simultaneously with a calculator as an explanation of the principles of decimal notation in reading and writing numbers.
- It gives a different view of regrouping (carrying, borrowing).

EXAMPLE 2

99 + 1 (see Illustration 37). Move [1]. [1] crossed the start line; therefore advance [10]. [10] crossed the start line; therefore advance [100] (Illustration 38).

As with any other computing device, the Abax should extend, but not replace, mental computation. Thus we recommend that children be told that they are expected to memorize addition facts, and that they are provided adequate training in this skill.

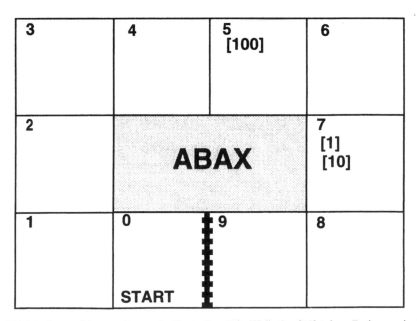

ILLUSTRATION 34. Start with 507 (shown in Illustration 33). Walk the [10] token 7 places clockwise to get 577.

3
[1]
4
5
[100]
6
2
ABAX
7
[10]
1
0
START
9
8

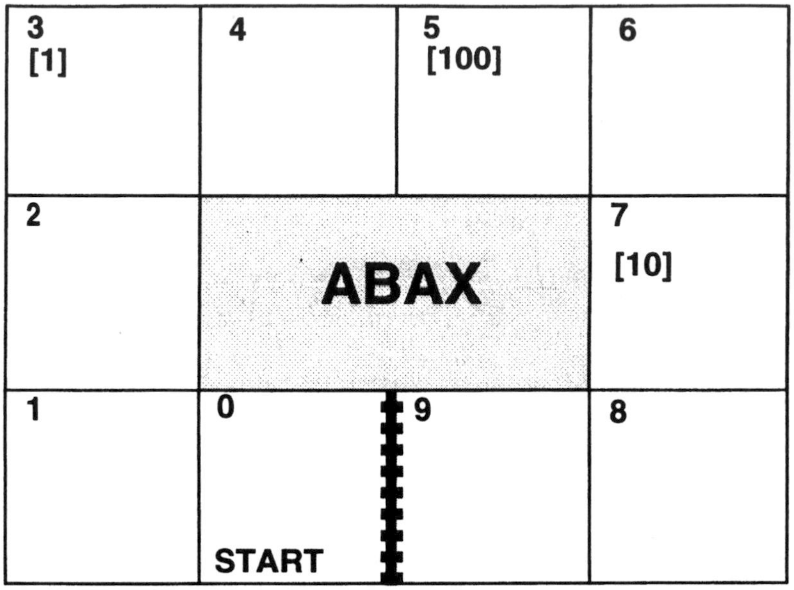

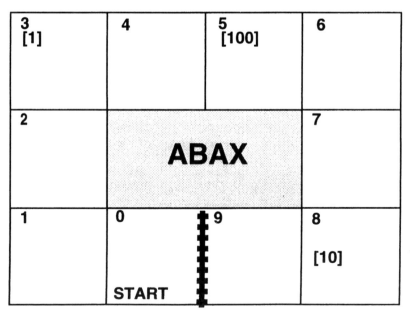

ILLUSTRATION 35. Start with 577 (Illustration 34). Walk the [1] token 7 places clockwise to get 573.

3
[1]
4
5
[100]
6
2
ABAX
7
1
0
START
9
8
[10]

ILLUSTRATION 36. Because the [1] token crossed the start line, advance the NEXT HIGHER token, [10], one place, giving 583.

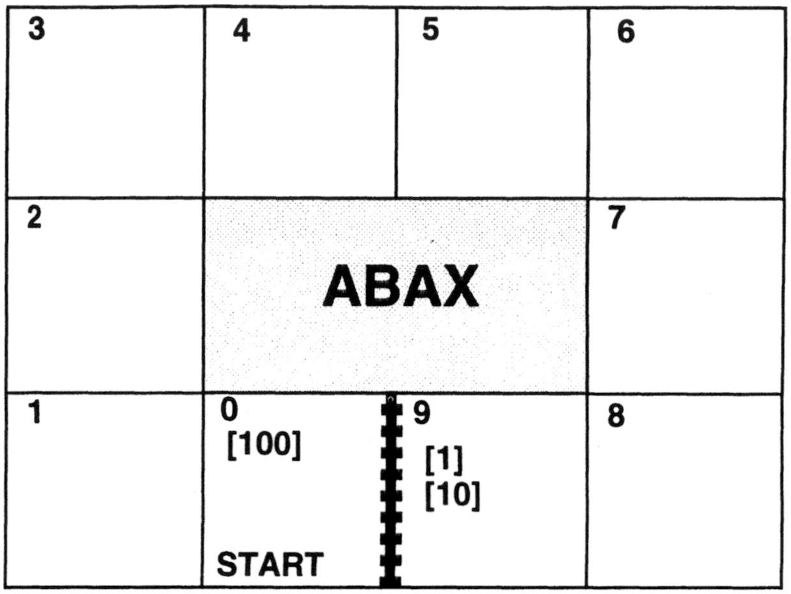

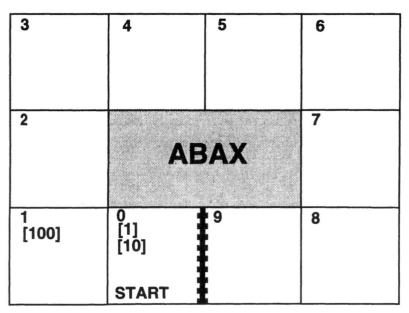

ILLUSTRATION 37. Represent 99. We are going to add 1 to it.

ILLUSTRATION 38. When we move [1], it crosses the start line. So we advance [10]. [10] crosses the start line, so we advance [100].

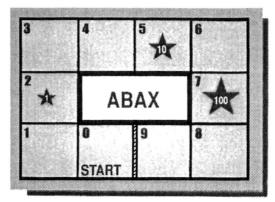

ILLUSTRATION 39.

Part II. How Abax was taught in Fourth Grade

When I walked into the fourth grade classroom, the teacher had on the blackboard: "Objective:" and "Supplies needed:" She asked me to fill in the blanks, as that's the way she starts all her lessons. So I wrote, "Objective: to learn to add in a fun way.
Supplies needed: your brain, a "counting board" (an ABAX; see board at end of lesson), one or more wooden stars, colored markers." The kids were giggling about the brain part, so I asked them how I would know if they had their brains. This turned into a fun discussion: "We're alive!"
"Well, a tree is alive. Does it have a brain?"
"Uh oh. Well, we can walk."
"A robot can walk. Does it have a brain?"
"No, it has a computer chip, and that's not a brain. We can talk."
"A radio can talk."
"Yes, but humans talk first, and then you hear it on the radio. Wait a minute, we can think!!"
"Well, that will be one way to show me you have brains: if you think during this lesson!"
    "The next thing we need," I said, "is a counting board, or an Abax. Have you ever heard the word Abax?"
    "No."
I wrote *abacus* on the board. "It is related to the word *abacus*. Do you know what an abacus is?" Several children had seen one, and they said it was a frame with beads on it that was used to do calculations. "That's right," I said. "And *Abax* is an old Greek word, meaning counting board. The ideas behind what we are going to do today are very old, so we have given the board an old name. On this board you will be able to do addition in a fun and simple way." I passed out to each child a board, printed on 8-1/2 by 11 inch heavy white card stock, which looked the one shown at the end of the lesson.
    I also put a transparency of the board on an overhead projector. "You can see that the board is just black and white. So we will decorate it with colored markers. Do you want to decorate it first, or would you like to play with it first to see what it does, and then decorate it?" The children voted to play first. "Okay, another thing we need is, hmmm, one or more wooden stars (see Illustration 39)."
    Beforehand, I had prepared 3 little bags of wooden stars (which can be purchased at craft stores): little stars with "1" written on them, middle-sized ones with "10," and big

ones with "100." I also had made stars out of a transparency, to use with the overhead Abax.

I gave each child a little star. "What do you see written on the star?" I asked.

"A one!"

"And what words do you see written on the board?"

"Abax, and Start."

"So let's put our little star on Start, or zero." (I did this on the overhead.) "Do you know how much our star is worth when it is on zero?"

"It is worth zero!"

"Good," I said. "Now add two, or move your star up 2." (Each time, I moved my star on the overhead.) "What is its value now?" Some children thought the star would be worth 3 when it was on 2 (1 from the star plus 2 from the board makes three, they reasoned). But others said, no, it was worth only 2 (2 from the board times 1 from the star). We agreed that it was worth 2 times 1. "Next let's add three. Where is our star?"

"On five."

"How much is 2 + 3?"

"Five."

"Good! Now let's see if we can add six. What is five plus six?"

"Eleven. And we are in trouble. We do not have eleven on our board," noted several children. The children counted up six from five, and their stars were on "1" on the board.

"Hmmm," I said. "What are we going to do? How can we get 11 on our board?"

It took a while, but finally Liliana said, "We need another star."

"Oh," I said. "I see here under 'Supplies needed' on the board that we need 'one or more' stars."

"I get it," said Brian. "We need a star that is worth 10."

"Gee," I said, "I am lucky! I just happen to have a bag of bigger stars, each marked '10.'"

I passed them out.

The children were quite excited. They called the little star the baby, and the bigger one the mother. The stars are thick and can stand upright on two points, so some children lined up their standing stars.

"Now that we have a '10' star, how can we make 11?" I asked. "Put the 10 star and the 1 star on one," several responded.

"Why does that make 11?

"Because 10 plus one is eleven."

"Great," I said. "Now let's add seven more."

"That's easy; we just move our one-star up seven, to eight." "So how much do we have now?"

"18!"

"Good! Now let's add six more." We moved our one-star up six, so it landed on four. "Wait a minute," I said. "What is 18 plus 6?"

"Twenty-four."

"And how much do we have on our board?"

"14."

"But look," I said. "When we moved our 1-star up six, we passed the heavy black line (between the 9 and the 0) on our board. Whenever your star passes the heavy black line, you have to move the next bigger star up one! So we have to move the 10-star from one to two. Do you have your brains? Do you get it?"

"Oh, yeah! It's like when we carry!" said Eva.

"That's right," I said.

"We're on 24 now. What shall we add next?"

"Forty! Let's add forty," said Stephanie and David.

"Piece of cake," I said. "How do you do it?"

"Simple; just move the 10-star up four places, to six."

"So what number do we have?"

"64!"

"Now let's add 29," I suggested. "Do you want to add the 20 first, or the 9 first?" "The twenty!" We moved the 10-star up 2, to 8.

"That gives 84."

"Okay, now let's move the 1-star up nine." "When we do that, it lands on three, but it passes 'Go,' so we have to move the 10 up one to 90." "Then what is the total?" I asked.

"93."

"This is fun!" said Brenda.

"It is also simple!" I said. "And 'Go' is a good name for the heavy black line!"

"What shall we add next?" Brian was already worried. "We can't add much more," he said, "unless we get another star!"

"Let's see what happens when we add 7," I said. "The one star lands on zero, but it passes Go. So we move up the 10-star to 0. But we need one hundred, and our board says zero! Maybe Brian is right," I said. "Do you think we need another star?"

"Yes!"

"How much should it be worth?"

"100!"

"We are lucky again," I said. "I happen to have some." I passed out large stars, marked 100.

"Now we have a daddy star!" said some.

"Yes, and we could go on and on. What would be the next star we would need?" I asked.

"1000, and then 10,000," said T.J.

"That's right," I said. "But I'm afraid this is the biggest star I have. How do we make 100?" "We put the 100-star on 1, and the other two stars on zero."

By this time the children were having no difficulty adding. "Let's make the number 248 on our board," I said. "It is 200 plus 40 plus 8."

"That is easy to do," said Jeremy. "Put 100 on 2, 10 on 4 and 1 on 8."

"Now let's add 379." I wrote on the board:

$$248$$
$$+379$$

"Shall we start with hundreds?" I asked.

"Sure," the children responded.

"Okay, 200 + 300 is 500." We moved our 100-stars up three. "Now tens. Forty plus seventy is . . . the 10-star goes up seven and lands on one. But it passed Go, so we have to move the 100-star up to 600. And now the ones. Eight plus nine is. . . . Well, there are two ways I could do this. I could do it in the old way, adding nine to eight, but here is a new way. If I want to add nine, it is the same as adding 10 and subtracting one, since nine equals 10 − 1. So I could move the 10-star up one and the 1-star back one. Then the 10-star is on two and the 1-star is on seven."

"Oh," said the teacher, "we have been learning about complements to 10!"

"Yes, we can add by using complements. So the answer is 627."

"What's the biggest number we can make with our three stars?"

"999."

"How does it look?"

"All three stars are on nine: nine hundred plus ninety plus nine."

"OK. Let's make 8 again." The 1-star goes on 8, and the 10- and 100-stars go on zero. "Now let's add 3." The 1 star goes on 1, but it passes go, so the 10-star goes on 1, making 11. "Oh, excuse me," I said. "I made an error. Would you please go back to where we started? Would you please take the three away from the eleven?"

"What?" asked several. "How do you do subtraction?"

"Oh," I said. "Eleven minus three: the 1-star goes back three, from 1 to 8. But it passes go (going backwards!) so I have to decrease the 10-star by one, from one to zero. So what is the answer?"

"Eight! Hey, we can do subtraction on this board too."

On the blackboard, under "Objectives," where I had written "To learn to do addition in a fun way," I wrote "and subtraction." "Well, let's check, and see if we can do subtraction," I said. I wrote

$$\begin{array}{r} 400 \\ -245 \\ \hline \end{array}$$

"Make 400, and subtract 245." This was a bit tricky for the children. Ana tried to compute with paper and pencil and got 265. I decided to go through the problem slowly with the children. "To start, we put 100 on 4, and 10 and 1 on zero. We move the 1-star back 5, to 5. But it passes Go, so we move 10 back 1, to nine. But 10 passes Go, so we move 100 back to three. Now we move 10 back 4, from nine to 5. And we move 100 back two, from three to 1. So the answer is one hundred fifty-five." The light began to dawn for most children.

"Let's try another hard one," said several. So I wrote:

$$\begin{array}{r} 601 \\ -378 \\ \hline \end{array}$$

"This time, you do it for me," I said. "If you can do it, I will know you have brought your brains along!" Eventually, all the children were able to do the problem using the Abax: Start by putting 100 on six, 10 on zero, and 1 on one. Move 1 back 8, to three; it passes Go, so move 10 back one, from 0 to 9. Ten passes Go, so move 100 back one, from six to five. Now move 10 back 7, from 9 to 2. And move 100 back 3, from five to two. The answer: 223.

"Whew," I said. "You brought your brains! Let's take a break now and color our Abaxes. Then we can work a few more problems, if you want, before we quit."

Colored markers were handed out, and the children decorated their boards in beautiful bright colors. Many made dots to represent the numbers, and they wrote their names on the backs of their boards. I told them I was going to leave the stars with their teacher, so they could play with the Abax after I was gone; they were delighted about this. We voted to keep the coloring time to 10 minutes, so we could try some more problems. Margarita asked if you can do multiplication and division with an Abax. I said I didn't know how to do it, but that you can add and subtract decimals with it, if you have pieces with decimal values.

I had brought a sheet of paper on which were a lot of addition and subtraction problems, and the children were eager to try some "hard" ones! The children reminded me that the biggest number we could get with our stars was 999, so I had to be careful in what I gave them. I first copied from the paper

```
   28
   35
   47
   63
  118
  400
```

We put 28 on our boards, and then added 35, followed by 47, 63, 118, and 400. The answer: 691.

Next we tried

```
   47
   56
  185
   -2
  349
  -88
  279
```

The children were not bothered by the minus signs in front of some of the numbers; they said, "We'll just subtract when we get to them!" This time, rather than working addend by addend, we elected to work with the tens column first; I drew an arrow pointing to it on the board, so we could keep our place. We put all stars on zero, and then moved our 10-star to 4, then to nine, and then to 4 (passing Go). We then moved the 100-star to one. We continued in this fashion, finishing the tens. We then worked with hundreds, and finally with ones. It took a while, but eventually everyone got 826 on the Abax. Some children said they didn't know that you could add (and subtract!) in so many different ways.

We ended with a subtraction problem:

```
  407
 -138
```

Some worked on it independently, and others worked in groups. Soon the answer was agreed upon: 269.

As we wound down, Brian said, "Boy, I could do math all day!"

"It is fun to make addition and subtraction into a game," said several.

The teacher and I collected the stars, boards, and markers.

SOME REFLECTIONS ON THE LESSON

I had not expected that the children would be so interested in this lesson, which is really one in learning to add and subtract whole numbers using a simple computing device. But they did not look at it as drill. They viewed it as Mathematics with a capital M; for most fourth graders, Mathematics means computation. They asked for harder and harder problems, and they did not want to stop when I had to go. The Abax seemed to take away their fear of difficult addition and subtraction problems; they knew that with it, they could handle anything I would give them. They began to play with numbers, breaking them into their expanded forms. They took real pleasure in adding or subtracting in non-standard orders, for example, taking the tens first, followed by the hundreds, and then the ones. The flexibility with which they experimented was wonderful to see. Some began using complements (e.g., adding 10 and subtracting 2, rather than adding 8; and even subtracting 10 and adding 1, rather than subtracting 9). Indeed, they brought their brains to the lesson.

Part III: A Game with an Abax

BASIC RULES

The game can be played by two to four players. Each player gets two tokens, [1] and [10], of the same color. (Different players have tokens of different colors.) Players start on 0; they take turns and throw two dice and advance their tokens by the TOTAL (2 to 12) shown, using rules of addition on the abax.

GOAL

The goal is to reach the number 100 (or more). The first person to do so is the winner.

ADDITIONAL RULES

(1) A penalty is assigned if a player gets the number 11, 22, 33, . . . , or 99. (This means that both tokens are on the same non-zero square.)
(2) A penalty is also assigned for an intrusion, when the [1]-token of player A lands on a square occupied by the [10]-token of player B.
(3) For the penalty, player A throws one die and subtracts (moves backward) its value from his or her position.
(4) Penalties are cumulative. If a player's [1] token lands on a square occupied by two [10]-tokens, the player takes two penalties consecutively. If, during the execution of a penalty, another penalty is acquired, it also must be executed.
(5) There is no penalty if the [10] token of an opponent is on square 0 (see the premium below).
(6) A premium is assigned when a player gets 10, 20, . . . , or 90 (his or her [1] token lands on 0). For the premium, the player throws one die and adds the value to his or her position.

3	4	5	6
2	ABAX		7
1	0	9	8
	START		

CHAPTER 12

Counting by Multiplication

I have taught this lesson in four different first grade classrooms. In each case, the teacher was also present.

What we plan to teach: Different ways to count the number of small squares which are inside rectangles (see handouts at end of chapter). One-to-one correspondence between number of pennies in a rectangle and number of small squares in the rectangle.

PROPS

For each child: 4 sheets of paper, 2 are square inch graph paper; on one are drawn 3 rectangles, 6″ × 4″, 3″ × 2″, and 5″ × 3″; they are labeled A, B, and C (see the first handout at the end of the lesson); on the other are drawn 3 rectangles, 8″ × 3″, 2″ × 2″, 2″ × 7″, labeled D, E, and F (see the second handout at the end of the lesson). The third sheet is square centimeter graph paper, with two rectangles, G is 18 cm × 6 cm, and H is 10 cm × 23 cm (see the third handout). The fourth is inch dot (not graph) paper (see end of book for the inch dot paper), containing only dots.

Also needed: 24 pennies (we handed them out in paper cups), pencil, calculator.

For the teacher: Transparencies of each of the four pictures to use with an overhead projector; overhead calculator; 24 pennies in a cup, transparency marker. A few extra pennies to replace ones that get 'lost'.

THE LESSON

"You should have 4 sheets of paper, a cup with some pennies in it, a pencil, and a calculator. Let's get out the first sheet. I will put mine here on the overhead projector. Do you see the letters A, B, and C? Can you tell me the names of the figures that are labeled A, B, and C?" Several kids knew they were rectangles. We had a brief discussion about how many sides and how many corners rectangles have, and how they are different from triangles. "OK, are you ready for the first problem? Look at rectangle A. We want to know how many small squares are inside it." The children said there are lots. "Let me tell you how we are going to find out exactly how many there are. We will take pennies from our cup, and put one penny on each square, and then what will we do? We will count the pennies." The children were very excited to have real coins. The class was pretty noisy as they dumped out their coins onto their desks and began to place them on the squares. It was a wonderful task. We talked about whether you have to go straight across or whether you can put your pennies down in any crazy old order. We decided order didn't matter. The desks were arranged in small groups, and children within a group would make different patterns with their pennies as they filled their rectangles.

"Do you think you have enough pennies in your cup to put one in each square of rectangle A?" The children did not think so. "Try it and see." I began to put pennies into each square in my rectangle A, using the transparency.

"How many squares are there? How many pennies do you need?" (Several children were able to count and say, "24.") "Do you have enough pennies?" The children said they had just exactly enough. I had to replace a few "lost" pennies that had disappeared by rolling off desks.

"How do you know there are 24 squares? How did you count them?" Most children simply counted 1, 2, 3, 4, ... "Let's see if we can figure out some quicker ways to count the number of squares. Do you know what a 'row' is?" Most children said that they knew. I showed on my transparency, using a pen, what a row is.

"How many rows of pennies do we have in rectangle A?" The children counted with me, and we determined there are six. "And how many pennies are in each row?" Again we counted, and we got four. "Can you figure out a way to count the number of squares, using your calculator? We have how many rows?" (six) "And how many in each row?" (four) One or two children said we could add up the number of pennies in each row. I asked how many times we would have to add four, and I wrote on the board as they described:

[4][+][4][+][4][+][4][+][4][+][4][=]

"Try it, and see what you get!" After most children had gotten the answer, I demonstrated it on the overhead calculator. "Let me show you a short cut. It is this (I pressed the keys and said aloud): [4][+][=][=][=][=][=][=]. Try it and see!" We waited until all children got the program to work. The children were surprised that it worked!

"Now, can we figure out some more different ways to count our squares?" (No one suggested either counting by columns or computing 4 times 6.) "Do you know what a column is?" The children did not know. The teacher said, "A column is like a row, but it goes up and down instead of across." I showed, using my marker and transparency, the columns in rectangle A. I asked, "How many columns of pennies are in rectangle A?" The children knew there were 4 columns. "And how many pennies are in each column?"

"Six!"

"So can we figure out another way to do it on our calculator?"

Several children began, "Six plus. . . ."

I wrote on the board:

[6][+][6][+][6][+][6][=]

"Try it and let me know what you get!"

"Now I will show you a shortcut again. Try this: [6][+][=][=][=][=]. What do you get?" Again we waited until all children tried the program.

"Can anyone figure out still another way to get the number of squares in rectangle A? I will give you a hint. There is a way to do it without using the plus key. Any ideas?" The kids did not know. "How many pennies are there in a row?

"4."

"And how many in a column?"

"6."

"Do you know what this key ([*]) is?" (The children did not know its name.) "It is the 'times' key. It is for multiplication. Try this: [4][*][6][=]. We say, 'four times six.'

What did you get?" (The children tried this and found it pretty amazing! A few needed help finding the [*] key; they pressed [+] instead.) "Let's write '24' on rectangle A."

"Let's put our pennies back in the cup. And now let's go to rectangle B. We want to know how many small squares it contains. Can you put a penny on each square in rectangle B?" I did this along with the children. Some children put pennies outside the border of rectangle B; I walked around to help them put them only inside. "How many pennies did you put down?"

"6."

"Can you show me some ways to get 6 on the calculator?" (One child suggested, [1][+][=][=][=][=][=][=]. It works!)

"How many rows?"

"2."

"How many pennies in each row?"

"3."

"So how can we do it using the plus key?" Some children suggested [3][+][3][=]. "Try it!" "Now, who can tell me how to do it when we look at columns? Who remembers what a column is?" They remembered that a column is an 'up-and-down' row. And one child suggested [2][+][2][+][2][=].

"OK, who knows how to do it using the times key?" I erased the 4 and 6 from the previous problem, leaving [ ][*][ ][=]. The children were able to fill it in: [2][*][3][=]. "Let's write '6' on rectangle B."

"How about rectangle C? How many squares? Put in your pennies, and let me know." The children were getting more proficient at putting pennies in squares, and they enjoyed doing it.

"Who can tell me some different ways to get the number of squares in rectangle C using my calculator? How many rows?"

"3."

"How many in each row?"

"5."

"How many columns?"

"5."

"How many in each column?"

"3."

The children suggested programs, and we tried them on the calculator:

    [5][+][5][+][5][=]
    [3][+][3][+][3][+][3][+][3][=]
    [5][*][3][=]

"Be sure to write '15' on rectangle C."

We continued this process for rectangles D, E, and F. Each time, we put one penny in each square, counted the number in a row and the number in a column, and did several different computations which gave the same answer, finally writing our answer on the rectangle. The children worked together in groups at this point, and a group would raise hands when they figured out how many squares a particular rectangle contained. They told me how they did it, and I wrote the program on the board.

Then we turned to handout 3. It has square centimeters (and not square inches), so we couldn't use our pennies, which wouldn't fit. "What are we going to do? We want to know how many squares are in rectangle G, and our pennies are too big."

A math whiz in the class said, "We don't need to use pennies. We can just count the number of squares in a row, and the number of squares in a column, and times them!"

So, using an overhead transparency and a marker, I asked the children to count along with me. I put a dot in each square in a row, as we counted. Similarly I put a dot in each square in a column, as we counted. Some children actually put marks on their sheets too, as we counted together. We discovered there were 18 squares in a row, and 6 in a column. "So what do we need to times?" Several children said, [18][*][6][=]. "Who can read that number?"

"One hundred eight!"

"Let's write '108' on rectangle G."

"Finally we have rectangle H. How many squares does it have?" Children knew we needed to count the number of squares in a row, and the number in a column, and multiply them. Again we counted and put a dot in each square in a row and in a column. Our program was [23][*][10][=]. Most children could read the number!

Then a kid said, "Hey, I wonder how many little squares there are on this whole sheet! Can we find out?" This created a good discussion. The final outcome was that we counted slowly and put dots into squares in a row and in a column of the whole sheet. The children figured out that our program needed to be [26][*][20][=]. There are 520 little squares on the whole sheet! This problem was a big hit for first graders.

We were out of time, so I told the children that tomorrow they could use their last sheet to make their own rectangles and count how many squares each contained. I also told them to look around the room. "Do you see any rectangles filled with squares?" They saw a page of a calendar, panes in the windows, and ceiling tiles. "Do you think you can figure out how many there are?" They told me they could!

We collected materials and ended the lesson.

What to do differently: At first there was some confusion about where the boundaries of the rectangles were on handouts 1 and 2; children made errors putting their pennies in squares that were 'inside' the rectangles. It might have been better for the borders to be colored. Also, if there had been more time, I would have asked children to calculate, for example, both [23][*][10][=] and [10][*][23][=], to show them explicitly that order doesn't matter.

B

C

A

D

F

E

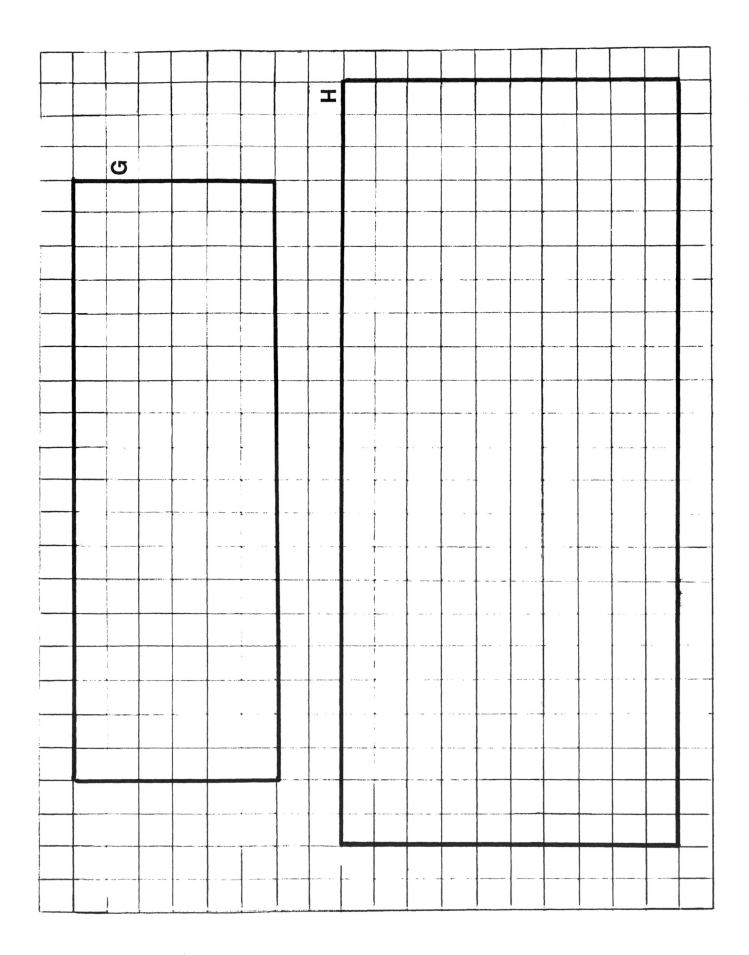

Examples:

[63][÷][3][=] shows 21., thus 63 = 3*21.
[63][÷][4][=] shows 15.75, thus 63 is not divisible by 4.

(3) The smaller of two whole divisors of a whole number is not bigger than the square root of this number. (This is a good place to learn about the square root in general, and also about the square root key on a calculator.)

Examples

One hundred can be built in four different ways:

$$100 = 2*50 = 4*25 = 5*20 = 10*10$$

Which numbers smaller than one hundred can be built in five or more ways? (Are there any such numbers? Yes, but finding all of them may take a long time.) Here is the complete list (5 ways is the most):

$$60 = 2*30 = 3*20 = 4*15 = 5*12 = 6*10$$
$$72 = 2*36 = 3*24 = 4*18 = 6*12 = 8*9$$
$$84 = 2*42 = 3*28 = 4*21 = 6*14 = 7*12$$
$$90 = 2*45 = 3*30 = 5*18 = 6*15 = 9*10$$
$$96 = 2*48 = 3*32 = 4*24 = 6*16 = 8*12$$

REMARKS

Notice that we consistently write "the result" on the left of the equality sign. In this lesson children should learn that = is used also as the sign of a relation between numbers, and not only as a command to do something.

From the very beginning the teacher should follow this pattern:

$$6 = 2*3 = 3*2$$ should be read, "6 is equal to 2 times 3 and is equal to 3 times 2."

Of course when children press the equal sign on the calculator, they give a command to do something.

EXAMPLE OF A CONVERSATION

Teacher: "How do you know that 60 is equal to 4*15?" (Here, *equal* expresses a relation.)

Student: "I pushed 60, divide, 4, and the equal button, and got 15." (Here, pressing the "equal" key is a command to perform an operation.)

Teacher: "What else do you know about 60?"

Student: "60 is equal to 2 times 30." (Now *equal* expresses a relation between 60 and 2 times 30.)

(The answer "2 times 30" would be unacceptable, "I know 2 times 30 about 60" is gibberish. The answer "2 times 30 is 60" is better, but not as good as the answer given above, "60 is equal to 2 times 30.")

CHAPTER 14

Counting Straws

This lesson is about counting by grouping, and is suitable for lower elementary grades. A proof of the number of ways a collection of ten objects may be counted is given below and may be suitable for middle school.

THE LESSON

In the lesson, computations are done mentally. Each child gets 10 straws. The task is to count the straws in a special way. You put all straws in front of you and pick up one or more at a time. As you pick up a group of straws, you must say the *total* number that you have in your hands and *not* the number of straws you are taking at the moment.

Examples

(1) You pick up one straw at a time and say, "One, two, three, four, five, six, seven, eight, nine, ten."
(2) You pick up two straws at a time and say, "Two, four, six, eight, ten."
(3) You pick up all ten straws and say, "Ten."
(4) You pick up 1, 2, 3, and 4 straws and say, "One, three, six, ten." (See Illustration 40.)

Children should take turns. And the teacher should write the name of a child and the numbers the child says on the blackboard. So after the four examples shown above we would have on the blackboard:

Mary	1 2 3 4 5 6 7 8 9 10
Josh	2 4 6 8 10
Anita	10
Sam	1 3 6 10

The rule is that the next child must choose a new pattern, so counting by ones, counting by twos, and picking up all the straws at once may not be repeated after Mary, Josh, and Anita have used them.

COMMENTS

You may change the number of straws that are counted, depending on the level of your class. The number should be at least 6, and values above 20 are not recommended.

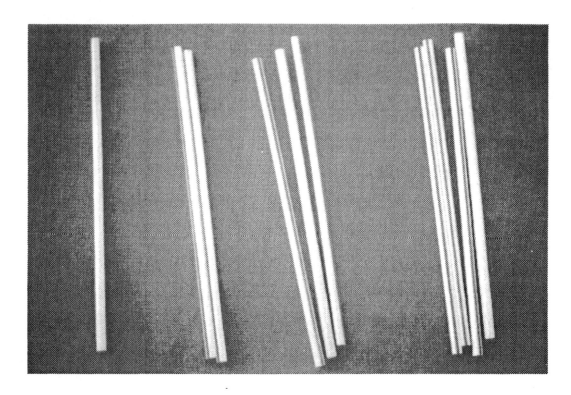

ILLUSTRATION 40. You pick up 1, 2, 3, and 4 straws and say, "One, three, six, ten."

Some Mathematics (for the teacher only or for middle school)

1. You may be interested to know in how many different ways a bunch of ten straws can be counted. The general formula is:

n straws can be counted in 2^{n-1} ways.

Number of Straws	Number of Ways	List of Ways
2	2	Two. One two.
3	4	Three. One three.
		Two three. One two three.
4	8	(Make the list yourself.)
5	16	
6	32	
7	64	
8	128	
9	256	
10	512	
11	1024	

So you see that you need at least six straws for a class of 20 children.

2. A proof that there are 512 ways that the 10 straws may be picked up is as follows (optional).

Let's represent one way of getting to 10 as follows:

s s s l s s s s s l s s

The s's represent straws, and the vertical bars indicate how many straws were taken (from left to right). So the above represents, take 3, then 5, then 2, and say, "Three, eight, ten."

Any pattern of bars, from none: s s s s s s s s s s (take 10) to all: s|s|s|s|s|s|s|s|s|s (take one straw at a time), shows a way to get to 10. How many patterns of bars are possible? There are nine positions between the straws. Each position either is empty or contains a bar (there are two choices per one position). Thus the total number of choices is $2*2*2*2*2*2*2*2*2 = 2^9 = 512$.

MONEY

CHAPTER 15

Buying Donuts

I taught this lesson in two first grade classes. Each child had a calculator, a pencil, and the donut ads (see handout at end of lesson). I used an overhead calculator and a transparency of the donuts ads on the overhead projector.

THE LESSON

"I am planning a party and I need your help. I want to buy 15 donuts for my party, and here are the ads from two donut shops. I want to buy my 15 donuts from the shop where the price is cheaper. At which shop should I buy, Donna's Donuts, or Donut Delight?"

The class was pretty silent at this point, so I suggested, "Maybe we should read the two ads first. Who would like to read the first ad?"

Erica read, "Donna's Donuts. Donuts thirty cents each. Buy 12 donuts and get the 13th free. Baker's dozen."

I asked the children what a dozen is. Julian said a dozen was 12. The teacher said that a baker's dozen was 13.

"Now who can read the second ad?

Justin read, "Donut Delight. Donuts thirty cents each. Buy 12 donuts and we charge you for only 11."

"Do you remember my question? I want to buy 15 donuts for my party. At which store can I buy them more cheaply?"

There was still silence. The teacher asked, "Does anyone have any idea where Dr. Pat should go to get her donuts? How can we help her?"

Susan suggested that, if one donut is 30 cents, we can figure out how much 12 cost. She said we could "times" 12 and 30. The teacher suggested we use our calculators here. "What keys do we need to press?" she asked. Emily said she thought it was [12][*][30][=]. The teacher reminded her that the 30 cents needed to be written with a decimal point. So I wrote on the board: [12][*][30][=] and asked the children where a decimal point was needed. Jimmy came up to the board and put in the decimal point. "Let's try it on our calculators. Can you find the times key, and the decimal point key?" I asked. This was the first time that some children had used the multiplication key, so I asked Jimmy to point it out on the overhead projector. The children tried [12][*][.30][=]. I asked them to read the answer (the display showed 3.6). They had some difficulty here; some read it as 36. Susan knew that wasn't right. If one donut cost 30 cents, then 2 cost 30 + 30 or 60 cents. So no way could 12 cost only 36 cents.

85

The teacher asked the children to look closely at the 3.6. "Does the number have a decimal point?" she asked. They agreed that there was a decimal point. "How can we read the number?" she asked.

Maggie said, "Three point six."

"That is good," said the teacher. "The three tells how many dollars, and the six tells how many dimes. Who can read it now?"

Justin said, "Three dollars and six dimes; that is three dollars and sixty cents!"

"How do you make 3.6 look more like three dollars and sixty cents?" asked the teacher.

"You put on a zero!" said several children.

So I wrote on the board, $3.60.

"Who can tell me what we have found here? What does three dollars and sixty cents mean? How many donuts can I buy?"

Susan said, "You can buy 12 donuts for $3.60."

"Okay, at which shop?"

The children said, "At Donna's Donuts."

Mike said, "Wait a minute. You can buy 12 donuts for $3.60, and then you get one free. So that makes 13 donuts. And you want 15 donuts. So you have to buy two more."

"Why do I have to buy two more?" I asked.

"Because 13 plus 2 is 15."

"Wow, that is right," I replied. "So how much more do I pay?"

Some children realized we had already discussed that. Susan had said, if one donut costs 30 cents, then 2 cost 60 cents. Mike said, "You need sixty cents more."

"What keys do I need to press? I have 3.6 on the display. What next?" "You need to add sixty more cents," replied several children. The teacher reminded them that you have to use a decimal point. I wrote on the board [3.60][+][60][=]. "Who can show me where the decimal point goes?" Again, a child came to the board and put in the decimal point.

"OK, let's try it. Do you have 3.6 on your display?" (The children did). "Now press (I read from the board) [+][.60][=]. What do you get?"

The children read, "Four point two." "What does it mean?" I asked.

Several children said, "Four dollars and twenty cents."

"Who can tell me what I get for $4.20?"

Mike knew: "You have to pay $4.20 for 15 donuts at Donna's Donuts."

"Okay, then, what about Donut Delight? How much do I pay for 15 donuts there?"

Again, there was silence. And again the teacher tried to get a discussion going. "Let's read again the Donut Delight ad. Vicki?" Vicki read the ad again.

Susan said, "If you get charged for only 11, then you need to times 30 by 11. But you get 12. So then you need to buy 3 more, because 12 plus 3 is 15."

"Okay, Susan, can you tell us how to use our calculators?"

She said, "Point thirty times 11. . ."

"And what, to make the calculator compute?" I asked.

"Equals," she replied.

I wrote on the board: [.30][*][11][=]. "Let's try it. What do you get?"

The children said, "Three point 3." Several knew that means three dollars and thirty cents.

"And how many more?" I asked.

There was agreement that I would have to pay for 3 more. "Each one costs 30 cents," said Justin.

"So how can I do it?" I asked.

"Just plus 30 three times," said Justin.

I wrote: [+][30][+][30][+][30][=]. The teacher asked, "Does she need a decimal point?" The class knew decimal points were needed, and Justin came to the board and placed them correctly. "Let's try it now. Do you all have 3.3 on your display? Now press [+][.30][+][.30][+][.[30][=]. What do you get?"

"Four point two! That is four dollars and twenty cents! It is the same at Donna's Donuts and at Donut Delight! So you can buy your donuts at either place. It doesn't matter!" This result created real excitement.

"Well, I have to go to one place or the other. Can anyone think of any reason why I might choose one place over the other, if the prices are the same?" Here the class got extremely creative. Some suggestions:

- Go to the store where the donuts have the most frosting.
- Go to the store that is open when you want to go.
- Go to the store that has the most chocolate donuts.
- Go to the store where the donuts are bigger or fresher.
- Does either store give away free balloons? Then go there.

It was time to stop the lesson. I thanked the kids for helping me figure out it didn't matter (at least price-wise!) where I go to buy my donuts and told them there was another question at the bottom of the ads, and they could work on it tomorrow. They kept their ads, and we collected the calculators.

What to do differently: If there had been more time, I would have shown them even more different ways to get to the two answers. (For example, for Donut Delight, [3.30][+][.30][=][=][=].)

Donna's Donuts

Donuts   30¢ each
Buy 12 donuts
and get the 13th free!
(Baker's Dozen)

Donut Delight

Donuts   30¢ each
Buy 12 donuts
and we charge you for only 11!

1. How much do we have to pay for 15 donuts at each shop?

2. Which is a better buy for 24 donuts: $3.30 for a dozen, or 27¢ each?

CHAPTER 16

Ali's Natural Foods

This lesson is suitable for grades three and up.

PROP

A handout of prices from Ali's Natural Foods Market (one per child). The handout is given at the end of the unit.

THE LESSON

Reading and Discussing the Handout

Children should read the handout, and the meaning of each line should be discussed. For example,

Garden of Eatin'		
Blue corn chips		
Reg. or no salt	15.5-oz.	1.69

means that a 15 1/2 ounce bag of blue corn chips costs $1.69, and that its brand name is "Garden of Eatin'."

What Are the Unit Prices of the Products?

The unit prices are prices computed either in dollars or in cents, of a unit amount of the product.

(1) What are the units used in the handout? Ounces (oz.), pounds (lb.), pints (pt.), tablets (tabs.), single napkins (the word each was shortened to ea.), single caramel apples, single vegetable rolls, boxes (of cereal), tubes (or bottles or cans) of hair care products.

(2) How are prices written? All are written in dollars. Example: .99 means $.99 which is 99 cents.

(3) How do we compute a unit price? We divide the total price by the amount. Example: The unit price of corn chips is:

   [1.69][/][15.5][=]   display: 0.1090322 dollars for one ounce.

(4) How do you read this display? 0.10 is $.10, which means 10 cents, but the rest, 0.0090322, means that it is more, almost 1 cent more. So the answer is: The corn chips cost $.11 (read 11 cents) for one ounce.

Remark:

Start with easy cases, and progress to more difficult ones, always rounding the result to a cent. Expressing the unit correctly, as in "11 cents per ounce," or "8 cents for one tablet," is absolutely necessary. Saying just 8, or 11, is meaningless, except as an answer to a direct question, such as, "How many cents per ounce . . .?"

(5) How do we compare unit prices?

You can compare unit prices only if the units are the same. You cannot compare dollars per pound to cents per napkin! Which prices can be compared, and which cannot?

Remark:

This is not as clear-cut as one may think. Can you compare the unit prices of apples and oranges, of caramel apples and vegetable rolls? This is a very nice topic for class discussion.

Special Tasks

EXAMPLES

Which item costs the most per ounce? Which the least? How to change units from dollars or cents per pound to dollars or cents per ounce? (Divide by 16, because there are 16 ounces in one pound.)

Unit prices for carrots,

$$[3.49][/][5][=] \quad 0.698 \qquad \text{(70 cents per pound)}$$
$$[/][16][=] \quad 0.043625 \quad \text{(4 cents per ounce)}$$

Remark: Do not convert between dollars and cents, because prices are written in dollars but are always read either in cents or in dollars and cents.

A General Remark

Intentional spelling errors such as "Schop" and "Eatin' " should be pointed out to the children.

GROCERY SHELVES

Shelton's
TURKEY CHILI
Mild or spicy — 15-oz. **1.59**

Barabra's
WHEATINE BITS
All natural — 6.5-oz. **.99**

Knudsen
CIDER & SPICE
A great Autumn beverage — 32-oz. **1.49**

Garden of Eatin'
BLUE CORN CHIPS
Reg or no salt — 10-oz. **1.69**

Enrico's
SALSA
Mild, hot, reg, no salt — 15.5-oz. **1.99**

Arrowhead Mills
BREAKFAST CEREAL FLAKES
Many varieties — **1.99**

BODY CARE

AUBREY
HAIR CARE PRODUCTS
All natural — 25% off

Glowing Touch
ALMOND SKIN CARE OILS
Assorted scents — 8-oz. **4.99**

DAIRY CASE

White Wave
SOY YOGURT
NO dairy- many flavors — 6-oz. **.79**

Land-O-Lakes
Butter
Salted quarters — 1-lb. **1.49**

HOUSEHOLD

Earthrite
LAUNDRY LIQUID
Environmentally sound — 64-oz. **4.99**

Green forest
Napkins
Environmentally sound — 250/ea **1.49**

FRESH MEAT CASE

The Pork Schop of Vermont
MAPLE SUGAR-CURED BACON
Nitrite-free — 1-lb. **2.49**

VITAMINS

Arbor Farms
ORANGE JUICE CEE
Chewable vitamin C — 150 tabs. **3.99**

Arbor Farms
HI POTENCY B-STRESS
Complete stress formula — 90 tabs. **6.99**

PRODUCE GARDEN

California
ORGANIC CARROTS
Fresh & crisp — 5-lb. bag **3.49**

Arbor Farms
CARAMEL APPLES
With dry-roasted peanuts — 2 for **.89**

Yoder
WHITE OR YELLOW POPCORN
Amish-grown — 2-lb. bag **.99**

FROZEN FOODS

Ben & Jerry's
ICE CREAM & FROZEN YOGURT
Many flavors — pt. **1.99**

Knudsen
RASPBERRY NECTAR
From concentrate — 12-oz. **1.99**

NATURAL DELI

Arbor Farms
VEGETABLE ROLLS
Ricotta & herb filling — **1.99**

Ali's Natural Foods Market
111 El Camino Real
Las Cruces, NM 88004
(505) 971-3990

CHAPTER 17

Shoe Sale

Fabulous Footwear on the mall had a sale on Reebok sport shoes.

Kind of Shoes	Regular Price	Sale Price
Little boys' Classic Nylon	28.00	19.99
Girls' Bangles	38.00	29.99
Women's Freestyle High	56.00	44.99
Women's Powertrainer	65.00	51.99
Men's D-Factor Mid	70.00	55.99
Men's Aurora	55.00	43.99

THE LESSON

There are two questions: "How much money do you save on each pair of shoes?" and "What percentage of the regular price do you save?"

(1) Children should first prepare data sheets. They may work in groups. The data sheet for a group may look like this:

Kind	Regular	Sale	Saving	Percent Saving

(2) They should decide how to compute the answer. The methods should be discussed. Proposed programs may be written on the blackboard.

Example:

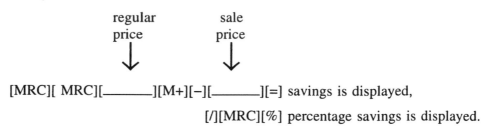

[MRC][MRC][_____][M+][−][_____][=] savings is displayed,

[/][MRC][%] percentage savings is displayed.

(3) Computations are performed. Tables are filled out. It is important that proper units (dollar sign and percent) be written on the data sheet. Values should be rounded to the nearest cent and to the nearest percent. The lesson may end in a discussion. Possible topics: sales, savings, good bargains, and so on.

Example of a partially filled out data sheet:

Kind	Regular	Sale	Saving	Percent Saving
Boys'	$28.00	$19.99	$8.01	29%
Girls'	$38.00	$29.99	$8.01	21%
Woman's Freestyle High	$56.00	$44.99	$11.01	20%

CHAPTER 18

Pay 'n' Pack

This lesson has been taught in several third grade classes.

PROPS

Each child gets a Pay 'n' Pack cash register receipt (see handout at end of lesson), a calculator, and a paper and pencil. Children work in groups of two or three. Each group gets a scale that measures in ounces and grams. The teacher has a bag of bananas, some pears, a few kiwi fruit (if available), some plums, and some kumquats (if available), to be handed out for weighing.

THE LESSON

Story

I picked up a few items at the grocery store yesterday afternoon. Here's the cash register receipt I got.

Task

Children read and discuss the cash register receipt, figure out what everything means (e.g., What is Reg 6? What does lb mean? What is a Bosc pear? What does *bulk* mean? What is a kumquat? What do F and T mean?), and check prices of the bananas, strawberries, pears, plums, and kumquats, using their calculators. For example, bananas cost 39 cents a pound, and I bought 0.81 lb.

[0.81][*][.39][=] display: 0.3159

Why was I charged 32 cents for my bananas? Stores round their prices up to the nearest penny.

What SLSTAX means and how it is computed are discussed.

Sales tax is charged only on non-food items in the state where I was shopping. So it is charged only on the dice, which cost $1.69. The sales tax was 10 cents. What percentage is the tax?

[.10][/][1.69][%]   Display: 5.9171597

So the tax rate is about 6%.

ILLUSTRATION 41. Weighing a pear.

Finally, each group gets a banana, a pear, a plum, and a kumquat to weigh. Children use scales to try to estimate how many bananas, pears, plums, and kumquats I bought (see Illustration 41). Since the scales weigh in ounces, and halves, quarters, and sixteenths of ounces, conversion from hundredths of an ounce to sixteenths is needed.

(e.g., [0.81][*][16][=] Display: 12.96

So I bought about 13 ounces of bananas.

Weights are recorded and discussed.

If kiwis are available, their approximate price per pound can be computed. After the lesson, children may eat the fruit!

PAY 'N' PACK

07/11/97   1:48 PM   FRI   REG 6 TERESA H

BANANAS
   0.81 LB @ 1LB / $.39   .32 F

STRAWBERRIES BULK
   0.89 LB @ 1LB / $.99   .88 F

BOSC PEARS
   0.58 LB @ 1LB / $.99   .57 F

KIWI FRUIT
   1 QTY $.28   .28 F

BULK PLUMS
   0.23 LB @ 1LB / $1.49   .34 F

KUMQUATS
   0.15 LB @ 1LB / $2.39   .36 F

DICE   1.69 T

    SLSTAX   .10

    TOTAL   4.54

    CASH   10.04

    CHANGE   5.50

CHAPTER 19

Selling Candy

This is a very old problem, suitable for upper elementary children.

THE LESSON

A store owner decided to mix three kinds of Halloween candies.
He had 10 lb. of red candies which he was selling for $1.99 per pound.
He had 8 lb. of green candies which he was selling for $2.49 per pound.
And he had 6 lb. of yellow candies which he was selling for $2.99 per pound. (see Illustration 42.)
How much he should charge for one pound of the mixture, to get the same amount of money that he would get if he sold them individually?

The key question is, how much did he expect to get for the candies? The answer is easy:

10*1.99 + 8*2.49 + 6*2.99 dollars

How many pounds of candy did he have? Another easy answer:

10 + 8 + 6 = 24 pounds

So if you divide these two numbers, you will know how much the store owner should charge for one pound of the mixture.

Computation

[10][*][1.99][M+]
[8][*][2.49][M+]
[6][*][2.99][+][MRC][/][24][=] gives 2.4066666.

So the store owner will probably sell the candies for $2.40 per pound, or $2.41 per pound, or (more likely) $2.39 or $2.49 per pound, because prices often end with 9.

ILLUSTRATION 42. Candy for sale.

CHAPTER 20

Prices in 1870

Around 1870 (over 125 years ago) the prices of everyday items were very different from what they are today (see Illustration 43). This does not mean that the items were cheaper, because people were also earning less.

THE LESSON

Following are some selected prices from 1870. Find out the current prices for the same items, and compare them with the old ones. Which prices increased the most, and which increased the least? Compute the percentage of increase.

Clothing:

- a coat $26
- a hat $5
- a pair of pants $9
- a pair of children's shoes $3
- a pair of boots $8

School Books:

- a geography $1.75
- a reader 85¢
- an arithmetic 65¢
- a speller 30¢

Food:

- a barrel of wheat flour $8, a barrel of apples $4, a barrel of salt $4.50
- a bushel of wheat $1.80, a bushel of corn $1.15, a bushel of beans $1.60, a bushel of potatoes 65¢
- a peck of apples 20¢
- a pound of beef 10¢, a pound of butter 37¢, a pound of sugar 18¢
- a pint of cooking oil 8¢
- one dozen eggs 18¢
- one quart of cherries 10¢, one pint of plums 4¢
- one orange 8¢, one banana 5¢

ILLUSTRATION 43. Prices in 1870.

Other Items:

- a watch $75
- a pencil 8¢
- a cord of wood $8

Can you name some products that were not available in 1870? (Cars, radios, computers, TVs, VCRs)

The units of measure that were used were slightly different from those used today. Products such as grain and fruit were sold by volume in units of "dry measure":

    2 pints are 1 quart
    8 quarts are 1 peck
    4 pecks are 1 bushel

Remarks: A bushel contains 2150.4 cubic inches.

In measuring grain, seeds, and small fruits, the measure must be level; but in measuring potatoes, apples, and other large articles, the measure must be heaping full.

Weight measures:

1 barrel is 196 pounds; 1 bushel is 56 pounds.

PUZZLES AND GAMES

CHAPTER 21

Birds and Cages

PROPS

Children work in groups of three. Each group gets a deck of 21 3 by 5 cards; each card represents a bird cage. On seven of the cards there are 10 dots each (each dot represents a bird). On seven cards there are five dots each (five birds). And seven have no dots at all (no birds).

THE LESSON

The task is to divide the birds and cages among the three children so that each child gets the same number of birds and the same number of cages. (This is a modified version of an old problem that can be found in chapter 8 (Seventh Heaven) of: Malba Tahan (1972; translation copyright 1993). *The Man Who Counted*. New York: W.W. Norton & Company.)

Two solutions are shown in Illustrations 44 and 45. The thin black lines divide the cages (cards) into three groups, each with seven cages and 35 birds.

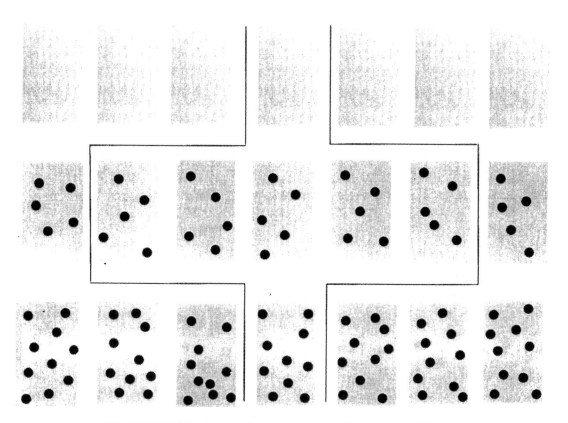

ILLUSTRATION 44. One solution for evenly dividing cages and birds.

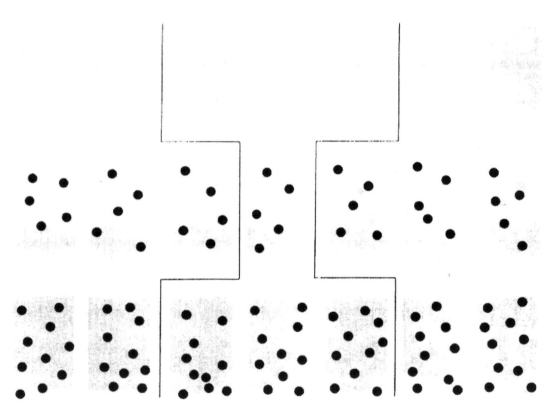

ILLUSTRATION 45. Another solution for evenly dividing cages and birds.

CHAPTER 22

Puzzle with Dominoes

Counting is often an essential part of mathematical reasoning. The following problem is a nice illustration. It involves tiling with rectangles and combinatorics. It is suitable for grades four through eight.

THE LESSON

Problem

Consider a 6 by 6 board (see handout at end of lesson). Cover it with domino pieces. (Each piece covers 2 squares.) The cover splits if it consists of two separate rectangles (Illustration 46). Does every cover split?

Activity

Children are given dominoes and paper with a square grid of dots (see grid at end of lesson). Each square is slightly bigger than half of a domino piece. (Each child needs 18 dominoes.)

(1) Each child draws a 6 by 6 square board with a pencil and a ruler. Neatness and precision count. At the very end of the lesson children will color and decorate their boards.
(2) The problem is explained and children try to find a cover that does not split by putting dominoes on their boards. This activity should last until most of them give up and start thinking that it is impossible.
(3) Now the question is, "How can we show that it is impossible?" Different proposals can be discussed. The teacher should at some point state the goal: We have to prove (do not avoid this word) that in every cover there is a line that does not cut any dominoes.
   Remark: Be sure that children understand what the word *cut* means in this context.
(4) A proof (to be shown by the teacher): Do not try to lead children to discover this proof. Give each child a written copy and go over it word by word, and sentence by sentence. Understanding each sentence does not guarantee that the whole reasoning is understood, but it is the main step toward it.

   • A line always cuts an even number of dominoes.
     There is an even number of squares on each side of a line. Thus if a line cuts some

105

ILLUSTRATION 46. A cover that splits into two rectangles. How it splits into two rectangles is shown on the right.

dominoes, the number of halves on·each side is even. If it cuts none, remember that zero is an even number.

• At most 9 lines can cut dominoes.

Each line that cuts a domino cuts at least 2 of them, as shown above. There are 18 dominoes on the board so at most 18/2 = 9 lines can cut some of them.

• There is a line that does not cut any dominoes.

There are 10 lines inside the board, five vertical and five horizontal. But we already proved that at most 9 lines can cut dominoes. Thus at least 10 − 9 = 1 line cuts none.

(5) The proof can be further discussed, and children should be asked to color and decorate their boards. This last activity is important.

This proof is not easy to understand, so many children will not get it. (Do not repeat it over and over again. It does not help!)

Coloring and decorating the boards will prevent the lesson from ending as a failure for these children.

CHAPTER 23

Number Game: Getting to One

This game is suitable for upper elementary grades.

THE LESSON

Choose a positive integer. Now, if it is even, divide it by 2; if it is odd, multiply it by 3 and add 1. When you reach 1, stop—you have reached the goal. (You can then play again by choosing a new starting number.)

Example 1

I choose 5. It is odd, so I compute $5*3 + 1 = 16$. Sixteen is even, so I compute $16/2 = 8$. Still even, $8/2 = 4$. Four is even, $4/2 = 2$. And again, $2/2 = 1$. Stop!

You can play this game mentally or with a calculator. You have to recognize whether the number on the display is even or odd. (If it is not an integer, you have made an error.) Then you must press either [/][2][=] (if the integer is even) or [*][3][+][1][=] (if the integer is odd).

Example 2

Choose 7. It is odd.

Press	Display
[7]	7.
[*][3][+][1][=]	22.
[/][2][=]	11.
[*][3][+][1][=]	34.
[/][2][=]	17.
[*][3][+][1][=]	52.
[/][2][=]	26.
[/][2][=]	13.
[*][3][+][1][=]	40.
[/][2][=]	20.
[/][2][=]	10.
[/][2][=]	5.

We saw 5 in the example above.

[*][3][+][1][=]	16.
[/][2][=]	8.
[/][2][=]	4.
[/][2][=]	2.
[/][2][=]	1.

Stop!

Try the number 27. For it you have to be very patient!

CHAPTER 24

A Basket Full of Eggs

THE LESSON

A young girl brought a small basket full of eggs to the market. Someone asked her, "How many eggs do you have?" She answered, "I do not know. But when I counted in fives, one egg was left. When I counted in sixes, five eggs were left. And when I counted in sevens, six eggs were left." How many eggs were in the basket?

Solution

Write three columns of positive integers:

- those that have remainder 1 when divided by 5
- those that have remainder 5 when divided by 6
- those that have remainder 6 when divided by 7

(Here, division means integer division with remainder, and not the usual division!)

Programs

For the first column: [1] [+][5][=] [=] [=] [=]...
For the second column: [5] [+][6][=] [=] [=] [=]...
For the third column: [6] [+][7][=] [=] [=] [=]...

Find the number (or numbers) which occurs in all three columns.

Results

1	5	6
6	11	13
11	17	20
16	23	27
21	29	34
26	35	41
31	41	48
36	47	55
41	53	62
46	59	69
...	...	...

Here we have found a possible answer: 41 eggs. Are there any others?

Yes, for example, 251. Can you find another one?

(Hint: Add 5*6*7 to the previous solution.)

Which of the numbers is a correct solution?

You cannot carry 200 or more eggs in a small basket. So 41 eggs is a correct answer.

CHAPTER 25

Digital Clock

This puzzle is appropriate for upper elementary and middle school.

THE LESSON

Yesterday I looked at my digital clock, and it was exactly 22 minutes after two (Illustration 47). I thought, "What percentage of the time are all the digits shown on my clock the same?" Can you help me solve this problem?

Solution

Here are the combinations of digits that I want to see:

        1:11
        2:22
        3:33
        4:44
        5:55
        11:11

Each one lasts one minute, and it occurs once in every 12-hour period. Therefore we have to compute 6/(12*60), and write it in percent form.

But this formula can be rewritten, and some calculations can be done mentally, so we have many different ways to make this computation.

Formulas	Programs on TI-108
6/(12*60)	[12][*][60][M+][6][/][MRC][%]
6/12/60	[6][/][12][/][60][%]
1/120	[120][/][%]

In each case the answer is 0.8333333%. But it is wrong to give the answer with seven digits after the decimal point. You do not have digits displayed all the time; changing the display is not instantaneous. Also no clock is accurate enough to justify such precision. Thus a more reasonable answer would be 0.83%. You know that the calculator shows 1/3 as 0.3333333, so you may also write .8 1/3%, which is read, "point, eight and one third, percent." The digits on a digital clock are all the same less than one percent of the time.

Do you want to see it happen? Set the alarm!

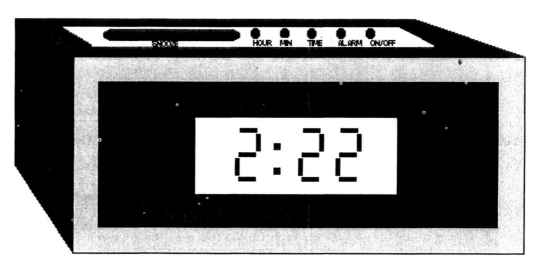

ILLUSTRATION 47. What percentage of the time are all the digits shown on a clock the same?.

CHAPTER 26

Old Shepherd's Will

This unit is suitable for upper elementary and middle school.

THE LESSON

Many problems which deal with sharing when the division is into unequal parts, lead to computing a sum of unit fractions such as:

$$1/2 + 1/3 + 1/6 + \ldots + \ldots = ?$$

You have probably heard the problem of the last will of an old shepherd:

"My oldest child will get 1/2 of my herd, the second oldest will get 1/4, the third oldest 1/8, and the youngest will have 1/16." The shepherd left 15 cows. How can we divide them?

The solution: Borrow one cow, making a herd of 16 cows. Now give 8 cows to the oldest child, 4 to the second, 2 to the third, and 1 to the youngest. 8 + 4 + 2 + 1 = 15, so we have one cow left, which we return to the lender.

Of course, this strange solution is due to a lack of arithmetical skills on the part of the old shepherd. 1/2 + 1/4 + 1/8 + 1/16 = 15/16, so he willed less than he had. (He willed only 15/16 of his herd.)

You can use different sets of numbers for similar problems.

Three good sets (for a shepherd with three children) are:

- 1/2 + 1/3 + 1/7 = 41/42 (herd of 41 cows only)
- 1/2 + 1/3 + 1/8 = 23/24 (flock of 23 sheep)
- 1/2 + 1/3 + 1/9 = 17/18 (herd of 17 camels)

CHAPTER 27

Smart Sally

PREREQUISITES

This lesson, suitable for upper elementary students, requires skill in paper folding. Some practice in origami during art classes should precede this lesson.

PROPS AND TOOLS

Rulers, glue sticks; squares of paper colored only on one side (origami or gift wrapping paper): 2 squares of red paper, 3 by 3 inches; one blue square, 5 by 5 inches; and 2 green squares, one 4 by 4 inches and the other 2 by 2 inches. Squares of white paper of corresponding sizes for practice in folding, etc.

REMARK

The lesson will work with squares of any sizes, but those above give interesting measurement problems.

STORY

Emily, Adam, and Sally each wanted to make a square that was the same color on both sides. Emily had two red squares that were 3 by 3 inches each, so she promptly glued them back to back. (Do not glue anything yet, but measure your red squares to check if they have the same size as Emily's.)

Adam had one blue square that was 5 by 5 inches. (Check the size of your blue square.) He also figured out how to make one smaller square, blue on both sides. Can you figure out how he did it? He did not cut his square at all! He just folded it and glued it.

Sally had the hardest task. She had two squares of different sizes, 4 by 4 inches and 2 by 2 inches. (Measure your green squares.) But she also managed to make one square that was green on both sides, without cutting any of her squares!

Task 1

Make a 2-sided red square like Emily's.

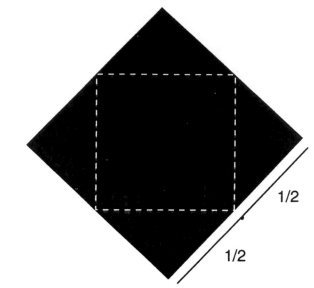

ILLUSTRATION 48. How to fold a square to make a smaller square.

Task 2

Figure out how to fold the blue square (as Adam did), and do it. (Fold the four corners so that they meet in the center of the square; see Illustration 48.)

Task 3

Figure out how to fold the big green square (as Sally did), and finish the task by gluing on the smaller green square (see Illustration 49; fold on the dotted lines.)

Task 4

Figure out the areas of the three two-sided squares, and figure out the lengths of their sides, and then measure the sides.

(1) **Red square:** No problem; a side is 3 inches long, and the area is 3*3 = 9 square inches.

(2) **Blue square:** The big blue square was 5*5 = 25 square inches. After it is folded, half of it is on one side and the other half is on the other side. So the area of one side is 12.5 (read: twelve and a half) square inches. So the length of a side is the square root of 12.5 inches.

> Keystrokes: [12.5][√]   display: 3.5355339

So a side is a little longer than 3 $\frac{1}{2}$ inches.

Remark: To get the answer in sixteenths of an inch:

subtract the integral part,   [−][3][=]   display: 0.5355339
multiply by 16,   [*][16][=]   display: 8.5685424

and round the result. The answer is $\frac{9}{16}$ of an inch.
So 3 $\frac{9}{16}$ inches is a little better (theoretical) estimate of the length of a side than 3 $\frac{1}{2}$ inches.

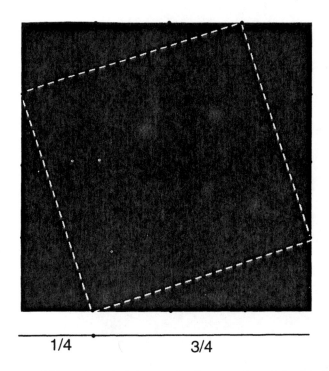

1/4   3/4

ILLUSTRATION 49. How to fold a 4 by 4 inch square, leaving room for a 2 by 2 inch square in the middle.

(3) **Green square:** The total area of the green square is $16 + 4 = 20$ square inches. So, 10 square inches of green are on each side. We have to compute the square root of 10,

      [10][√]   display: 3.1622776

and to convert the decimal fraction into parts of 16,

      [−][3][*][16][=]   display: 2.5964416

Answer: The length of a side is somewhere between $3 \frac{1}{8}$ and $3 \frac{3}{16}$ in. (because $3 \frac{2}{16} = 3 \frac{1}{8}$).

CHAPTER 28

Square Money

This lesson was taught in eighth grade. The teacher invented a country named after herself!

THE LESSON

Story

Our currency is generally available only in some denominations: 1¢, 5¢, 10¢, 25¢, 50¢, $1 = 100¢, $2 = 200¢ (have you seen a 2 dollar bill?), $5 = 500¢, $10 = 1000¢, $20 = 2000¢, $50 = 5000¢, and $100 = 10000¢. Let's imagine a country called Ur (you may invent another name) in which the unit of money was called tal (you may invent another name). All coins in Ur were square in shape, and the available denominations were just squares of whole numbers.

Here are the values of Ur's coins:

1t	4t	9t	16t	25t
36t	49t	64t	81t	100t
121t	144t	169t	196t	225t
256t	289t	324t	361t	400t

They did not need bigger denominations, because in the country of Ur nothing cost more than 400 tals.

Task 1 (for the Whole Class)

Design and make from poster board at least one full set of square coins for the city of Ur. You have to decide their sizes, pictures, and inscriptions.

One day a merchant announced, "No matter what purchase you make in my store, it can be paid for with at most four coins. If it cannot be paid as I say, you get it for free." So all the people tried to find a number of tals that cannot be paid in four or less coins, in order to get their purchase for free.

Task 2 (Individual or Group Work)

Can you find an amount (not more than 400 tals) that cannot be paid in four or less coins? (It must be a whole number.) Or maybe the merchant was right, and all prices up to 400 tals can be paid in four or less coins?

119

EXAMPLES

Price	How to Pay It
7t	4t + 1t + 1t + 1t,
400t	100t + 100t + 100t + 100t,
26t	25t + 1t, or 16t + 9t + 1t.
81t	81t

SOLUTION

The merchant was right. Each amount can be paid in four or less coins. This follows from a theorem proven by a French (Italian by birth) mathematician, Joseph Lagrange (1736–1813), which says that every whole number is a sum of four or less squares of whole numbers. (You count a square number as a sum of itself.) The proof itself is too difficult to be presented in school, but it can be found in most books about number theory.

MEASURING

CHAPTER 29

Circles and Angles

This lesson has been taught in two second grade classes. In both cases, the teacher was present, but the lesson was led by the instructor. The lesson can also be taught in third grade.

PROPS

For each child: a ruler, compass with well sharpened pencil (we placed the pencils in the compasses so that the pencils were about 1/4″ longer than the compass points), protractor (ours were clear plastic, with a diameter of six inches), pencil, 3 sheets of blank paper, 1 sheet with 5 angles (see the handout at the end of the chapter), calculator. A calculator poster was present in the classroom. We used an overhead projector, with a transparency of the handout. And plenty of extra blank paper was available. We also used an overhead calculator, but this is not necessary. Words written on the board: diameter, protractor, degrees, angles.

THE LESSON

Part 1: Introducing Circles

"Will you hold up your compass? Do you know what it is for? Have you ever used one before? Let's just take a few minutes and draw some circles. If you have any problems, I can show you how to do it." The instructor gave a demonstration, using her compass to draw on a piece of paper that was fastened to the blackboard. "Hold the compass at the very top. You might want to use only two fingers to hold it. Now stick the point into the center of the paper. That is going to be the center of your circle. Keep pressure on the point, and just swing the compass around, like this. See the circle? It's tricky to hold the compass so the point in the center doesn't slip. If yours doesn't turn out perfectly, don't worry. Just try again. We have plenty of paper for practicing. If you want help, raise your hand and we will come by and help. You can make as many circles as you want. Make two that are the same size. After that, you can make any sizes you want. Do you see how to get different sizes? You can just pull the two parts of the compass apart to make a larger circle, or push them together to make a smaller one."

After the children had played with the compasses for five or ten minutes, and each child had made at least one good circle, we continued.

123

"Now pick your best circle, and do this. Find its center. Can you see the hole your compass point made? Use your pencil and mark the hole with a dot. If you need help finding the center, let me know. Now take your ruler and draw a straight line across your circle, going through the center dot. Watch me; I'll do it here for the one I have drawn. Here is my center [she marked it with a dot], and now I am going to draw a line through it, from one side of the circle to the other, like this."

We waited until everyone had drawn a straight line.

"Do you know what a line through the center of a circle is called? I will write it on the board, and you see if you can say it." (She wrote *diameter*.)

Part 2: Introducing Protractors

"Now I am going to show you another way to draw a circle." (She held up her protractor.) "Do you know what this is?" No one knew its name. "I will write it on the board, and you can again see if you can say it." (She wrote *protractor*.) "What shape is it?" Kids said it was half of a circle. "What do you think it is used for?" They said, for drawing circles. "That is one thing we can do with it, but you will see something else we can do too. Before we use it, please take a new sheet of paper and draw a long straight line across it, something like this." She tacked up on the board a new sheet of paper, and using a ruler, she drew a long line, neither horizontal nor vertical, but skewed. "Now put a dot on it, somewhere near the middle of the line." She did so on her drawing. "Now we are going to use the protractor to draw a circle."

One student said, "Oh, I see what we are going to do!" He had taken his partner's protractor and put their straight sides together, forming a circle.

The instructor put her protractor on the overhead. It clearly showed a small hole at the zero point on the side with the straight edge. "See this hole? You need to put the hole exactly over the dot on your line, and line up the protractor so that it is straight with your line, like this." She showed how to do it, holding her protractor on the sheet tacked to the board. "Now you have to hold it there pretty tightly. And then you can draw around the protractor. Just trace around the half circle. Okay, now we have half. How will we get the other half?" Several children had already flipped their protractors over and begun to draw the other half. "Remember to put the hole over the dot, and line up your protractor with your straight line."

"Does everyone have a circle? If not, let us know, and we can give you some help."

"Now we are going to have some fun. We are going to be gremlins. And we are going to walk around the circle. Wendy, will you be the first gremlin? You can start your walk right here." (She marked an x on the circumference of the circle, on the drawn diameter, and on the left end of the protractor.) "I want you to take 90 steps. Each step is called a degree, and they are marked on the protractor. Can you see the degree marks? Do you see that not every one is numbered? They are numbered 10, 20, 30, and so on. Where will Wendy be when she has taken 90 steps? When you think you know, mark it on your circle with your pencil, and raise your hand."

We started at the left side of the protractor, because we wanted children to use the outer scale, which starts at 0 on the left, first. If we had started on the right, they would have needed to read the inner scale, which is a little harder to see.

Children were able to do this task. They marked their circles with a dot at 90 degrees. And they counted by tens to ninety. The instructor wrote *90 degrees* on the board.

Part 3: Measuring Angles

"Now let's get out the sheet with the drawings on it. Does everyone have the sheet? Do you know what the drawings are? I will write it on the board, and you see if you can say it." She wrote *angles*.) "Do you see that the angles are marked a, b, c, d, and e?

"Let's start with angle a. And we need a new gremlin. Andrew, will you be the gremlin? We need to mark Andrew's path and count the number of degrees he walks. It is a little tricky to draw his path. Watch me. I will do it on the overhead projector." A transparency had been prepared, that matches the handout. The instructor put the hole of her protractor over the vertex of angle a, and lined up her protractor so that the straight line from the hole to the left edge was on top of the "bottom" ray of the angle. Again, it was done this way to allow children to read the outer scale.

"See how I draw Andrew's path? He will walk from here to here." She drew an arc across angle a. "I will even draw an arrow, to show Andrew's direction. Now you draw Andrew's path. Do you need some help?

"Who knows how many steps Andrew took? How many degrees there are in angle a? When you know, raise your hand!" Almost all children were able to give the answer. However, one child responded, "25."

"Of what?"

"Millimeters!"

"No, they are not millimeters. They are parts of an angle. Who can remember what they are called?"

Another child: "Degrees!"

"Yes, degrees. So how far does Andrew walk?"

"25 degrees."

"Right. Let me show you the symbol that scientists use for degrees. It is °. So let's write on our angle, 25°." The children labeled angle a 25°.

We continued this process for angles b, c, d, and e: get a new gremlin, draw the arc of his or her path, measure the path in degrees, and label the angle. The instructor explained that you could use either side of the protractor, but that you have to be sure that you start counting from zero. (This is tricky, since zero is not shown as a numeral on the protractor.) Using the overhead, she pointed to the two scales, and she showed how you can measure from one side or the other.

Angles d and e caused some excitement. Angle d measures 4°, and the children had to label it with tiny writing, in order for their writing to fit inside the angle. They observed that the gremlin who walked 4° didn't walk very far! Angle e measures 135°, and it is the only angle they measured which was larger than 90°.

At the end, the instructor issued a challenge question: "Okay, let's go back to the circle we drew around the protractor. Who would like to be a gremlin now? Laura, how about you? Laura is going to walk all the way around the circle. She will start here, and end here. How many steps does she take? How many degrees are there in the whole circle?"

Children first said there are 170 degrees in a half circle (they thought this because 170 is the last number printed on the protractor). The instructor helped them to see that there are 10 more degrees, and 170 + 10 is 180. Several knew that you need to add 180 + 180 to get the amount in a whole circle. The instructor suggested they use their calculators. She used an overhead calculator and pressed the keys [180][+][180]. She invited Fred to come up and press the equals key. The children said, "360."

"360 of what?" asked the instructor.

"360 degrees."

"So how many degrees are in a whole circle?"

"360."

Materials were then collected.

The two teachers in whose class the lesson was taught plan to follow it up with more angle measurement, and we plan to do a lesson in which we look at and measure the angles in polygons (see for example Lesson 30, titled Quadrilateral). Also parts of the Circles lesson (Chapter 40) would fit nicely here.

Our protractors were 6 inches in diameter, and each of our angles had rays that were greater than 3 inches long. We plan to give children some angles whose rays are less than 3 inches, to see if they can measure them.

What we would do differently: The lesson was long and could have been better managed in two sessions. The children had difficulty lining up the rays of their angles with the line on the protractor, which was not along the protractor's edge, but was set in about $\frac{1}{4}$ inch. Also, the line on the protractor actually had a break in it (it was not continuous). The teachers suggested that this "inner" line could be marked more clearly, with a marker, to help children see it better.

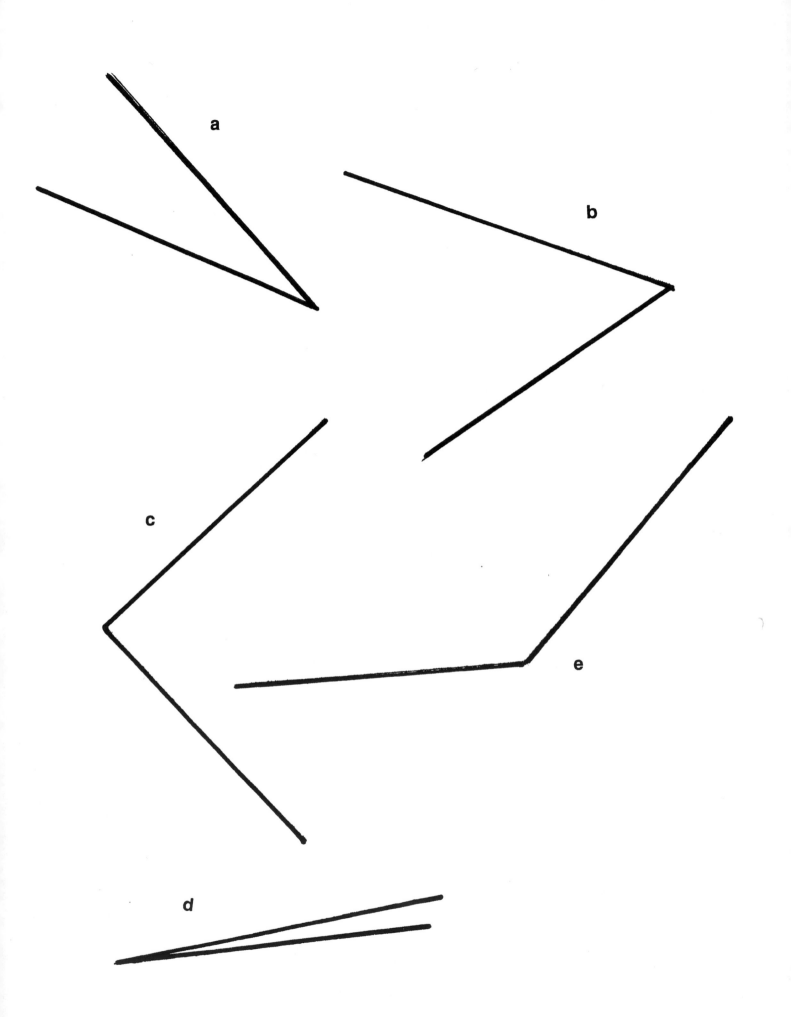

CHAPTER 30

Quadrilateral

This lesson has been taught in two second grade classrooms. Both a guest instructor and the teacher were present. The lesson was led by the instructor. The children had been taught the Circles and angles lesson the week before, and we recommend it as a prerequisite to this one.

PROPS FOR EACH CHILD

Two copies of a picture of a quadrilateral (see handout at end of lesson). Its vertices are labeled a, b, c, and d. Angles a and b measure 60 degrees; angle c is 150 degrees, and angle d is 90 degrees. Side ab is 8 inches, bc is 4 inches, cd is 3.5 inches, and da is 6 inches. Each child also should have a ruler, a pencil, a protractor, and a calculator.

OTHER PROPS

Calculator poster, 2 overhead transparencies of the picture, overhead calculator. (Overhead projection is helpful but not necessary.)

THE LESSON

Part 1: Getting to Know the Figure

"Let's check if we have everything we need. Do you have a ruler? Hold it up! What about this?" The instructor held up a protractor. "What is it? Who remembers its name?"

The children could not remember. They guessed "diameter" and "angle."

"Which would it be, between these two: compass or protractor?"

"Protractor!"

"Do you have two copies of the picture? And a pencil? And a calculator?" The instructor held up the picture, which was also showing on the wall via the overhead projector. "Do you know what this is?"

Various children said, "A kite! A diamond! A triangle!"

"No, it's not a triangle. Why not?"

"A triangle has three sides."

"How many sides does this thing have?"

"Four."

"So it's not a triangle. What is it? Is it a square?"

"No!"

"Why not?"

"A square has 4 sides alike."

"Is it a rectangle?"

"No!"

"How many corners does it have?"

"Four."

"So it has how many sides, and how many corners?"

"Four of each."

"I am going to write its name on the board. It's a long word that scientists use. It starts with a q. "Then the next letter is a u."

The instructor wrote in large letters *quadrilateral.* "Who can read it? Let's start with the first 4 letters. Can you say them?"

"Quad."

"And the rest?" The kids struggled here.

"When two babies are born together, what are they called?"

"Twins."

"When three babies are born together, what are they called?"

"Triplets."

"When four babies are born together, what are they called?"

A few kids answered, "Quadruplets."

"What did you say, for four kids? Quad? What's the name of this figure? Why do you think it starts with quad?" A few kids got this, but it was hard for them to explain: quad means four.

Part 2: Measuring Angles

"So we have its name: quadrilateral. We are going to find out a few things about our quadrilateral. How many angles does it have?"

"Four."

"The first thing we want to know is, how big is each angle? Do you see how each one is labeled, a, b, c, and d? Let's start with a. How big is a? How are we going to find out?"

"We use this" (protractor).

"OK, let's measure angle a. Do you remember how we did it last week? Remember that we were gremlins, and we walked along the paths we made?" She showed how to do it, using the overhead projector, with a transparency of the handout. "We have to put the hole in the protractor over the corner of the angle, and line up the side of the quadrilateral with the line on the protractor, like this." (Children still had some difficulty here.) "Now let's draw the path." Using a transparency marker, she drew an arc for angle a. "How far do we go? Do you remember how we measure angles?"

"We go 60."

"Sixty what? What are we measuring?" Only one or two children remembered that the unit we measure in is degrees. "OK, let's label angle a: 60 degrees. Do you remember how scientists write degrees? They don't write d-e-g-r-e-e-s. They use a special symbol. What is it?" Children remembered the small raised open circle. "So let's write 60°." (She wrote 60° on the transparency.)

"Now let's measure angle b. Can you line up your protractor and do it? Let me know if you need help. How big is it?" Many children were able to measure it, and called out, "It is sixty degrees too!" "OK, let's label angle b: 60°" (she wrote it on her transparency).

"We have two more angles to go. How big is angle c? Measure it! Is it bigger or smaller than angle b?"

"Bigger!" Children were able to measure it, and raised their hands. One kid said, "Angle c is 150 degrees."

"And angle d. How big is it?" Kids drew the arc and measured. It is 90 degrees. "Do you notice anything special about angle d? Take your other piece of paper, and see if you can fit one of its corners into angle d. Does it fit? We call this a square corner. How many degrees are there in a square corner?" The kids had no difficulty here. They responded, "90 degrees."

"OK, do we have all the angles measured and labeled? Now, here is the challenge question. How many degrees total do we have in our quadrilateral? Who can say? Who knows what we need to do?" The children knew that we need to add. "Let's use our calculators. What shall we add?" Some kids helped here. The instructor wrote on the board: 60 + 60 + 150 + 90 =. She asked children to try it on their calculators. Soon she began the computation on the overhead calculator. This was the first time the children had seen an overhead calculator, and they were excited: they wanted to know how it works. The instructor asked a kid to come up and press the final [=] on it. "What is the total? Who can read it?"

"360."

"360 what?"

"360 degrees."

"So what do we know about the angles in this quadrilateral?"

"When you add them up, you get 360 degrees."

"Do you remember anything else about 360 degrees? Think about last week, when we made circles." Only one or two children remembered that there are 360 degrees in a circle.

Part 3: Measuring the Perimeter

"OK, now we have a new task. Let's start again with our other picture. It's just like the first one, but there are no pencil marks on it. We want to know how long each side of our quadrilateral is. How can we find out?" The kids knew they needed to measure, using a ruler. They asked whether to use centimeters or inches. "Today we are going to use inches. Let's start with side ab. How long is it? Raise your hand when you know." The children had an easy time here: the side is 8 inches long. "Let's write on the line, 8 in. Scientists abbreviate inches in." (She wrote on her transparency.)

"How about side bc? How long is it?" Kids measured and found it was 4 in. long. "OK, let's write 4 in. on side bc. How about cd? It is a little tricky." Kids measured and found it was 3 and a half inches. "How are we going to write that?" One child came to the board and wrote 3 $\frac{1}{2}$. "OK, let's label side cd 3 $\frac{1}{2}$ in. Now how long is side da?" Kids measured and found it was 6 inches. "Let's label side da 6 in. Now here is another challenge question: how far is it around the quadrilateral? Do you know what the distance around something is called? I don't think you have heard the word before. It is called perimeter." (She wrote *perimeter* on the board.) "We want to find the perimeter of our quadrilateral. How do we do it?" The children knew we need to add the 4 numbers.

"So we need to add 8 and 4 and 3 $\frac{1}{2}$ and 6. OK, I will write the keystrokes:

[8][+][4][+]

"Do we have a problem? How are we going to enter 3 $\frac{1}{2}$?" The kids observed that there is no $\frac{1}{2}$ key! "Do you know how the calculator shows $\frac{1}{2}$?" A few children said, "Point five."

The teacher asked, "Why is .5 the same as $\frac{1}{2}$? Do you know that .5 means $\frac{5}{10}$, and $\frac{5}{10}$ is $\frac{1}{2}$? So the keystrokes are:

$$[8][+][4][+][3.5][+][6][=]$$

"What do you get? Who can read the number: 21.5? 21.5 what?" The kids replied, "Inches."

"What's another way to say twenty-one point five?"

Several kids responded, "Twenty-one and one-half."

The instructor had a yard stick and asked several children to come up and point to 21.5 inches on it. This was difficult; they first pointed to 25.

"That's all we have time for today. So let's collect all the supplies. But before we stop, who can tell me the name of this figure again?"

"Quadrilateral!"

Things We Might Do Differently

The lesson was long and could be broken into two parts, one about angles and one about perimeter. Other things we could have done, time permitting: Have children draw a diagonal of the quadrilateral and measure it. They could also then measure the angles in the two resulting triangles and sum them (the angles in a triangle sum to 180 degrees).

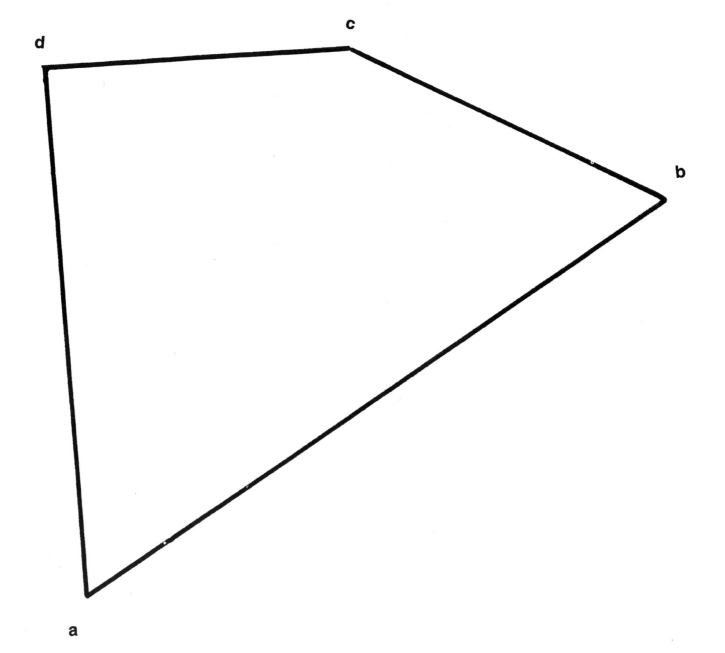

CHAPTER 31

Buttons

This lesson is suitable for middle elementary grades. Sorting buttons by color, shape, number of holes, etc, can be done in early elementary grades.

PROPS

Lots of buttons of different sizes, rulers, calculators, cups, and plates, labeling stickers. We prepared an overhead with diameters and sizes given below, and kept it in view for the entire lesson.

THE LESSON

"I have here a jar of buttons. I want them to be sorted by size. Here is the list of sizes" (see Illustration 50):

Diameter	Size
Less than 1 cm	size 1
Greater than or equal to 1 cm but less than 1.5 cm	size 2
Greater than or equal to 1.5 cm but less than 2 cm	size 3
Greater than or equal to 2 cm but less than 2.5 cm	size 4
Greater than or equal to 2.5 cm but less than 3 cm	size 5
Greater than or equal to 3 cm but less than 3.5 cm	size 6
Greater than or equal to 3.5 cm but less than 4 cm	size 7
Greater than or equal to 4 cm	size 8

Organization of Work (Example for 24 Children)

Children work in pairs; three pairs form a team, so we have four teams.

Team 1 consists of John and Mary, Samantha and Maria, and Pablo and Kim. Each pair gets a cup of buttons to sort, and smaller cups to keep them in.

(1) Each pair of children:

- labels the cups with appropriate stickers (e.g., size 3; see Illustration 50)
- sorts the buttons (measuring the diameter with a ruler)
- prepares and fills out a data sheet

131

ILLUSTRATION 50. Sorting buttons by size.

The following is an example:

Pair 1: Pablo and Kim

Size		Number
1		0
2	‖	2
3	‖‖‖ ‖‖	8
4	‖	1
5	‖‖‖ ‖‖‖ ‖	11
6	‖	1
7		0
8	‖	2
		Total 25

Now children put their buttons into big cups prepared and labeled by the teacher. *Remark.* It is important that all buttons are safely on the teacher's desk before the next part of the task.

(2) Each team gets together and prepares their team's data sheet.
An example follows:

| | Team 1 | | | |
Size	John and Mary	Samantha and Maria	Pablo and Kim	Sum
1	1	0	0	1
2	2	0	2	4
3	6	12	8	26
4	3	5	1	9
5	8	12	11	31
6	2	3	1	6
7	1	1	0	2
8	0	0	2	2
Total	23	33	25	81

Next, each team chooses two representatives who finish the task.

(3) The eight representatives prepare the final report for the teacher. An example of a heading is shown:

| | Number of Buttons of Different Sizes | | | |
Size	Team 1	Team 2	Team 3	Sum
1	1	2	0	3
.	.	.	.	.
.	.	.	.	.
.	.	.	.	.

IMPORTANT REMARK

The number of categories and the diameters for each category that are used here are just an example. They should be adjusted depending on the distribution of sizes of the collection of buttons used in this lesson.

This activity can be followed by dividing buttons of one size into subcategories, for example, by number of holes.

CHAPTER 32

How to Measure the Value of Pi

Pi is the ratio of a perimeter of a circle to its diameter. Pi is an irrational number (not a mixed number), so in all practical applications we use only its approximations. When people were using mainly common fractions, 3 $\frac{1}{7}$ was the most popular. Now 3.14 is the most common. In many everyday applications, even plain old 3 is good enough. But how to get any of these? Does one really need mathematics, or one can simply measure pi?

PROPS

Poster board, ruler, compass, round toothpick or paper clip, pencil, cash register tape, and cellophane tape.

THE LESSON

The task is to measure pi.

Method (see Illustration 51)

Draw a circle on stiff thick paper or poster board. Draw one radius. Measure it very carefully. Cut out the circle. Using the sharp point of your compass, make a hole in the middle. From a round toothpick or a straightened paper clip, make an axle and put it through the hole. Test if you can easily roll the circle holding the axle with both hands. Now draw a straight line on a (long) piece of paper. Tape down one end of the paper.

For the rest, you need a helper, because one person needs both hands to roll the circle, and the other person must watch and make a mark on the paper.

Start with putting the circle on the line. The radius drawn on the circle should be vertical and should touch the line. Let your helper mark this point on the line. Now roll the circle a full turn along the line, until the drawn radius is again in vertical position. This is the other point on the line that should be marked.

Measure the distance between the points marked on the line. This is the perimeter of your circle. Divide it by twice the length of the radius. This is your value for pi.

Example

Two measurements with one (very precisely cut out) circle. Radius = 4.2 cm (diameter 8.4 cm).

First trial: Perimeter = 26.5 cm
[26.5][/][8.4][=] returns 3.1547619

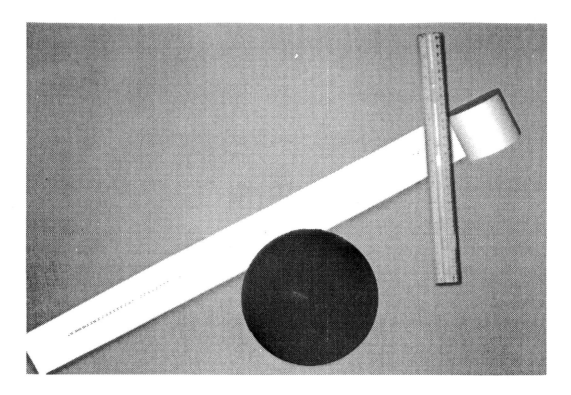

ILLUSTRATION 51. Materials for measuring pi.

Second trial: Perimeter = 26.4 cm
[26.4][/][8.4][=] returns 3.1428571

Comment

Physical objects, such as a wheel, do not have exactly the same properties as abstract objects such as circles. Thus a "crude" measurement may, in this particular case, better describe the reality than the "mathematically correct," theoretical answer.

More about Pi

(1) Just as all equilateral triangles are similar, so are all circles on a plane similar. This means that the ratio of the circumference of a circle to the diameter is always the same, independent of the size of the circle.

(2) What is this ratio, which has been labeled pi (the Greek letter π)? For centuries people tried to find out. Practically, you may use different approximations, 3, 3 1/7, 3.14, ... and so on. But what is the true value?

(3) We know now that π is irrational, just as $\sqrt{2}$ is irrational. This means that π cannot be written exactly as a mixed number. But there are efficient methods of getting as many digits as you want. As a challenge, people using computers have calculated millions and millions of digits of π.

(4) Recently I checked on the World Wide Web, and found more digits of π than I needed to know. Here are some:

3. 141592653589793238462643383279502884197169399375105820974944
592307816406286208998628034825342117067982148086513282306647
093844609550582231725359408128481117450284102701938521105559
644622948954930381964428810975665933446128475648233786783165
271201909145648566923460348610454326648213393607260249141273
724587006606315588174881520920962829254091715364367892590360
011330530548820466521384146951941511609433057270365759591953
092186117381932611793105118548074462379962749567351885752724
891227938183011949129833673362440656643086021394946395224737
190702179860943702770539217176293176752384674818467669405132
000568127145263560827785771342757789609173637178721468440901
224953430146549585371050792279689258923542019956112129021960
864034418159813629774771309960518707211349999998372978049951
059731732816096318595024459455346908302642522308253344685035
261931188171010003137838752886587533208381420617177669147303
598253490428755468731159562863882353787593751957781857780532
171226806613001927876611195909216420198938095257201065485863
278865936153381827968230301952035301852968995773622599413891
249721775283479131515574857242454150695950829533116861727855
889075098381754637464939319255060400927701671139009848824012
858361603563707660104710181942955596198946767837449448255379
774726847104047534646208046684259069491293313677028989152104
752162056966024058038150193511253382430035587640247496473263
914199272604269922796782354781636009341721641219924586315030
286182974555706749838505494588586926995690927210797509302955
321165344987202755960236480665499119881834797753566369807426
542527862551818417574672890977772793800081647060016145249192
173217214772350141441973568548161361157352552133475741849468
. . .

SCIENCE

CHAPTER 33

A Class Project about Temperature

This lesson is suitable for early elementary grades.

PROP

An outdoor thermometer located in a shady area is needed (see Illustration 52).

THE LESSON

Introduction

Many scientific projects start with a question. Then data are gathered, recorded, and analyzed, and the question is either answered or not. Only a few of many scientific investigations end with success.

Question

At what time of day when we are in school is it the coolest, and when is it warmest?

Method of Answering This Question

First we will gather the data. We will measure the temperature outside, before school starts, during the school day, and after school is out, for 5 days during one week. Then we will see if we can answer our question.

Organization

We have to prepare a chart, saying who is going to take a measurement and at what time. We need 5 daily charts. In order to avoid errors, two students will be responsible for each measurement.

Example 1

Monday, March 24, 1997

Period	Time	Temperature	Comments	Who Is Responsible
Before school	7:56	64°	cloudy	Adam, Bill K.
After 1st break	8:52	71°	sunny	Maria Q., Sue
After 2nd break				Josh, Pablo

. . . .

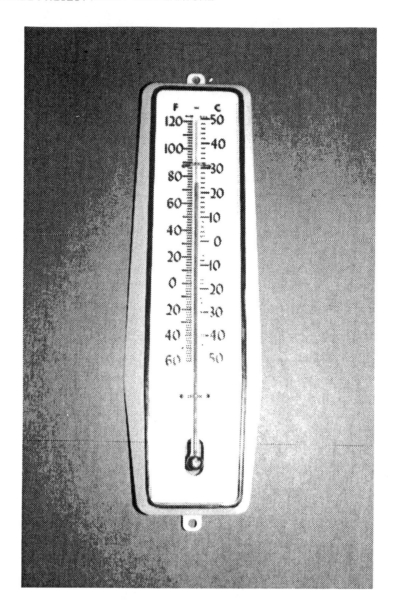

ILLUSTRATION 52. An outdoor thermometer to use for a class project.

The children who are responsible for the measurements must be decided on ahead of time, and their names should be written on the chart. The exact time a measurement is taken should be recorded. The teacher keeps the filled out charts in her desk.

During the week after the measurements are taken, the data are analyzed. The class is divided into 5 teams, each responsible for one day. Each team prepares a graph of daily temperatures and formulates a conclusion about the time of lowest and highest temperatures. The team should also compute the average temperature for its day.

Example of Conclusions

The coldest was clearly before school started, at 7:56. But later the temperature was going up and down, probably because of moving clouds, so we do not know when it was warmest.

After all the teams present their findings, the data are tabulated.

Example 2

	Lowest Time: Temperature	Highest Time: Temperature	Average	Comments
Monday:				
Tuesday:				
. . . . . .				

The final step is to answer the original question. Whether one can do it depends on the data. The pattern may be clear: coolest in the morning; hottest between 1 and 2 P.M. Sometimes no pattern may be able to be detected; the weather could have been very unstable. This could lead to other questions that may become the subject of the next study.

REMARK

The mathematical aspect of the project may seem small, but it is really quite important. Children are accustomed that numbers come from worksheets, and that they have to perform an operation on them and show the teacher the result. Here, numbers come from measurements, and we want to detect a meaningful pattern.

CHAPTER 34

Measuring Temperature

ABOUT THERMOMETERS

Thermometers are available in a large range of prices ($2 to $30) and quality. The ones we used cost about $50 for a set of 30. For classroom experiments, a temperature range between 20°F and 200°F is quite satisfactory. The minimum accuracy is 2°F (or 1°C), but it is better to have more sensitive instruments. Thermometer response time should be high; an "instant reading" is preferable. Because temperature measurements are usually a part of group projects, four or five thermometers for a classroom will suffice. The measurements below were done with a thermometer with an accuracy of 2°F; therefore all measurements are even numbers. Beakers are nice to use in this lesson, but they are not needed.

PROPS

For each team of four or five children, you need a thermometer, a watch, a cup or bowl to keep water in, and a cup for carrying hot water and ice.

THE LESSON

Start with cold water; add some hot water (from the faucet) to it, and stir; later, add some ice cubes and stir. Measure the temperature at one minute intervals. See what happens (see Illustration 53.)

Organization

Children work in groups of at least four.

- One child checks the time.
- One child reads the thermometer.
- One child writes down the time and temperature.
- One child handles the ice and hot water, and stirs the mixture.

Following is an example of a data sheet prepared by a team ahead of time and filled out during the experiment:

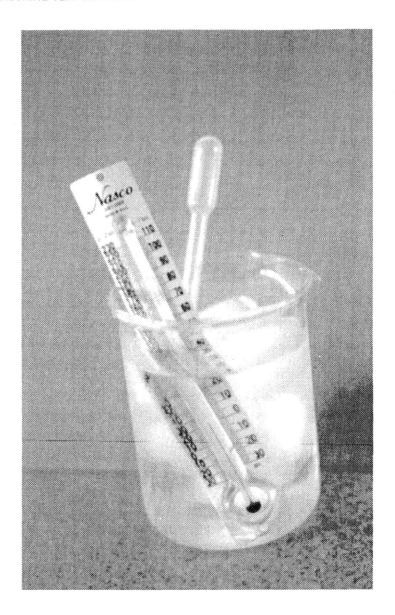

ILLUSTRATION 53. Measuring the temperature of ice water.

Names of Children:

Time in Minutes	Temperature in Degrees Fahrenheit	Remarks
9:20	46	Cold water
1	46	
2	48	
3	48	Hot water added
4	88	
5	86	
6	84	
7	82	Ice cubes added

Time in Minutes	Temperature in Degrees Fahrenheit	Remarks
8	66	
9	56	
30	54	
1	52	The ice melted
2	54	
3	54	
4	54	
5	56	
9:36		End of experiment

An example of a data analysis sheet prepared by this group for discussion by the whole class follows:

Names of Children:

Temperature in Degrees Fahrenheit	Rate of Change in Degrees/Minute	Comments
46		
	0	Cold water slowly warms up.
46		
	2	
48		
	0	
48		Adding hot water makes
	40	the mixture warm.
88		
	−2	
86		
	−2	Water is cooling slowly.
84		
	−2	
82		
	−16	Adding ice makes the cooling
66		very fast.
	−10	
56		
	−2	
54		
	−2	
52		After all the ice melts, the water
	2	starts slowly warming up again.
54		
	0	
54		
	0	
54		
	2	
56		

An interesting variant of this lesson is to use a lot of ice cubes until the water is near 32 degrees, and then to add salt to the ice water. This lowers the freezing point. An eye dropper of fresh (not salty) water can be put into the beaker of salty water, and the fresh water in the dropper will turn to ice, while the surrounding (salty) water remains liquid!

CHAPTER 35

Weighing

INTRODUCTION

Many math and science topics require weighing. Recently there have appeared on the market very cheap (about $5), but adequate, spring scales, so a class can be supplied with a set (one scale for two or three children). The material below has been used in grades kindergarten through eight. It assumes that children work in groups of two to four, and that each group has scales on their table. (The scales we used here were Good Cook scales, with a double, circular face weighing up to 18 oz and 500 g (see Illustration 54). The resolution was 1/4 oz, and 10 g. The tension of the spring was adjustable by a simple screw.)

THE LESSON

Preliminary Activities

These activities can be done in kindergarten. Scales are a tool, and children must learn to use them properly.

(1) Children choose small objects, books, rocks, plates, and so on and weigh them. Pay attention to correct reading of the weight and using appropriate units. (2 3/4 ounces, or 160 grams). Saying just the number without units is meaningless!
(2) Guess and check (use modeling clay): Children make a set of balls from modeling clay. They try to guess their weight (do not forget units), and check their guesses. This can be arranged as a contest.
(3) Small heavy and big light objects: Create a collection of small heavy objects (screws and bolts, and other iron objects), medium size medium weight objects (rocks, heavy wood), and light large objects (light wood, balsa).

Weighing Pennies

How much does one penny weigh? A solution: weigh 50 pennies (count by grouping), divide the total by 50 (answer: approximately 2.5 g).

What is the accuracy? On our scales the error can be close to 10 g for 50 pennies. So for one penny it can be close to $10/50 = 0.2$ g.

Remark: Many versions of this activity can be carried out. For example, how much

ILLUSTRATION 54. How much does one walnut weigh?

does one Lima bean weigh? How much does one walnut weigh? (see Illustration 54). In each case an estimate of the error should be done in the same way as above.

Using Ounces

When weighing in ounces, conversion into decimals should be used. Example: 25 pennies weigh 2 $\frac{3}{4}$ oz. How much does one penny weigh?

[3][/][4][+][2][=]	shows 2.75 ounces
[/][25][=]	shows 0.11 ounces per penny
[*][16][=]	shows 1.76 sixteenths, which is approximately 2/16 = 1/8 oz.

Density

Use rocks or other rather heavy objects having a size such that they can fit comfortably in a plastic cup (see also Chapter 37). An example of an activity follows:

(1) Weigh a rock. (In our example, it weighed 230 g.)
(2) Put the tray from the scales on the table. Put a plastic cup on the tray. Fill it very carefully to the brim with water. Lower the rock into the cup. The water will spill into the tray. Lift the cup and weigh the water remaining on the tray. (In our case the weight was 90 g.)
(3) Compute the density. [230][/][90][=] shows 2.5555555
(4) Interpret the result. The density of this rock is approximately 2 $\frac{1}{2}$ (*not* 2 $\frac{1}{2}$ grams!). This means that the rock is two and one half times heavier than the same volume of water.

Remark: In this activity, understanding what is measured is crucial. The volume of the water spilled is the same as the volume of the rock. So we are comparing weights of two materials, keeping the volume the same. The ratio of two weights (both are measured in grams) is an abstract number.

Other Activities

After children are familiar with weighing and have scales available all the time, the question, "How much does it weigh?" can be asked often as a small part of many other activities.

CHAPTER 36

Weighing Rocks

In this lesson children are given the task of lining up a collection of rocks, one for each child in the classroom, from lightest to heaviest, on a table in the classroom. The lesson has been taught in several first grade classes.

In preparation for this lesson, I went on a hike (with a backpack) and collected about 30 rocks, varying in weight from about 1.5 ounces to slightly over a pound. The rocks varied in their appearance: their colors, smoothness, textures, etc. I collected rocks of granite, quartzite, feldspar, chert, basalt, etc. The collection was not quite "ordinary"— the rocks were definitely appealing to young children.

PROPS

Children work in pairs, and each pair has a scales (we used inexpensive spring scales which have both ounce and gram scales). Each child should also have a calculator, a pencil, and one half of an index card, for recording the rock's weight.

THE LESSON

Story

The teacher brings a bucket of rocks into the classroom and tells the children that it contains her (or his) rock collection. She wants to display the rocks in a long line (a "rock snake") from lightest to heaviest, and she needs their help.

Each child selects a rock from the bucket. We decided to weigh our rocks in ounces, so we used the outer scale (see Illustration 55). The needle should first be adjusted to show zero ounces (this can be done with the screw on the back). The teacher can demonstrate how to weigh a rock, by putting one on her scales and showing children how to interpret where the needle points. For example, the needle may point about halfway between three ounces and four ounces, meaning the rock weighs about 3 and one-half ounces. She can show how to write "3 $\frac{1}{2}$ oz" on the index card. We recorded weights in $\frac{1}{4}$ ounce increments, showing children how to decide what to record when the needle was not pointing exactly to a scale mark.

After children weigh their rocks and record the weight, the rocks should be lined up from lightest to heaviest. I asked if anyone had a rock that weighed two ounces or less. Those who did came forward, with their rocks, cards, and scales. Which rock goes first? This can be decided by ordering the numbers on the cards from smallest to largest. But

ILLUSTRATION 55. Lining up rocks from lightest to heaviest.

children may have another idea: Put each rock on a scale, and the lightest one moves the needle the least!

Rocks weighing two ounces or less were lined up. Next I asked for rocks weighing three ounces or less, and so on.

The "rock snake" will grow.

Children may ask, "How much do the rocks weigh altogether?" The weights from the cards can be recorded on the board, and the numbers added, by the teacher and by the children, for example,

- 1 ½ is entered as [1][/][2][+][1][M+]
- 5 ¾ is entered as [3][/][4][+][5][M+]

The total weight (in ounces) can be seen with [MRC].

Dividing the answer by 16 tells how many pounds of rocks are in the collection.

CHAPTER 37

What Is the Density of a Small Rock?

This unit is suitable for upper elementary and middle school.

Density is the mass (weight) divided by the volume. (In this material we do not discuss the relation between mass and weight.) Density is measured in grams per cubic centimeter. The density of water is approximately 1 gram per cubic cm.

PROPS

A beaker with milliliter marks (we used 400 mL beakers), scales that measure in grams, a calculator. Here we are using metric tools.

THE LESSON

(1) Weigh the rock.
(2) Fill the beaker half way up with water. Write down the volume. Put the rock in. Write down the volume again. The increase of the volume (the difference of the two measurements) is the volume of the rock.

Example

- weight 37 g (grams)
- volume of water 60 cubic cm
- volume of water and the rock 72 cubic cm

[72][−][60][M+][37][/][MRC][=] display: 3.0833333

Answer: The density is about 3.1 grams per one cubic centimeter (approximately three times the density of water).

Variation

What is the density of an apple?

CHAPTER 38

Nutrition Content of a Snack

This unit is suitable for grades four through eight.

THE LESSON

Each child gets a pouch of Farley's Fruit Snacks (see Illustration 56). Children read and have explained the nutritional information; then they eat the snack.

The front panel contains advertising that makes you want this product because it is tasty and good for you.

What It Says	The Conveyed Idea
Strawberry Fruit Snacks	Tasty
Fat Free	Good for you
Made with Real Fruit!	Tasty
Enriched with Vitamins C, E, and Beta Carotene.	Good for you
Contains 25% of the RDI of these vitamins	A lot of good stuff
NO PRESERVATIVES	Good for you

- RDI stands for Recommended Daily Intake; it is the total amount of a food product that nutritionists think one should eat daily.
- 25% of RDI means that one pouch contains 1/4 of the daily amount of these vitamins.
- Preservatives are chemicals added only in order to have the food stay fresh longer. Many people think that food without preservatives is healthier.

The back panel contains ingredients used in this product and the nutritional value of the food. The ingredients are Fruit (grapes, strawberries), Corn syrup, Sucrose, Gelatin, Corn starch, Citric acid, Ascorbic acid (Vitamin C), Alpha tocopherol acetate (Vitamin E), Beta carotene (Vitamin A), Natural & artificial flavor, Red 40. The ingredients are listed in order of their amount. They are used for different reasons: nutritional value, taste, texture, and color. Some natural ingredients such as citric acid (found naturally in citrus fruit such as lemons) also act as preservatives. Red 40 is just food coloring.

Nutrition facts for one pouch state that there are 80 calories. This is the amount of energy you get from it. The total number of calories that people use daily varies. It depends on whether you are growing or you are an adult, what you do, and many other things. Typically people use about 2000 calories daily.

157

ILLUSTRATION 56. Fruit snacks.

Total Fat 0g (0%DV)   No fat in this product
Sodium 15mg (1%DV)   Some salt
Total Carb. 19g (6%DV)   Carbohydrates = flour and other starches
Sugars 14g   Sweet
Protein 1g   Proteins are mostly in milk, eggs, and meat
Vitamin A (25%DV, 100%
  as beta carotene)
Vitamin C (25%DV)
Vitamin E (25% DV)

The amounts are measured in metric units: grams (g), kilograms (kg), and milligrams (mg).

    1 kg = 1000 g = 2.2 lb (pounds)
    1 mg = 1/1000 g

One box of Fruit Snacks contains 10 pouches and weighs 9 oz, or 255 grams.

Daily values, DV, is an approximate amount you may eat daily. So if 15 mg of salt is 1% DV, this means that you should not eat more than 100*15 mg (100% DV) of salt daily. But 1500 mg = 1.5 g, so this is the total daily salt intake suggested by nutritionists.

Just stating what is not there: there is not a significant source of calories from fat, saturated fat, cholesterol, dietary fiber, calcium, or iron.

Some Suggested Activities

(1) How much is 1 g (1 g of water has a volume of 1 cubic cm.)?
(2) What is the daily value (DV) for carbohydrates from a pouch of Fruit Snacks?
(3) Express the amounts not in grams, but in ounces. How many grams are equal to one ounce?
(4) How many individual strawberry-shaped fruit drops are in the pouch? (You will find that this varies! When we did this, it led to some descriptive statistics.)
(5) How many calories are in each fruit drop?
(6) How many fruit drops would you need to eat to get 2000 calories' worth?
(7) Read the label of another product.
(8) Discuss the food pyramid.

This unit is quite open-ended, but the finale of course is to eat the fruit snack!

CHAPTER 39

Comet

This material, suitable for middle school, is in four parts.

(1) Comets and their properties
(2) Ellipses and their properties
(3) Drawing an ellipse
(4) Kepler's law and interpreting what we know about comets on the drawing

THE LESSON

Comets and Their Properties

Most comets are part of the solar system because they go around the sun. Other objects in the solar system (planets and asteroids) move around the sun in the same direction, and their orbits are almost in the same plane. Comets also follow this pattern. Typically, their orbits are in the same plane as planets, and they move in the same direction around the sun as the planets. The main differences between comets and planets are that

(1) Comets move on very elongated orbits, whereas planets move in orbits that are very close to circular.
(2) Most comets are not very large objects, compared to planets (especially the big planets like Jupiter), so during their travel around the sun, comets sometimes come very close to the sun and they are almost scorched, but most of the time they are very far away from the sun in a deep freeze.

What does a comet consist of? Mostly rocks and ice. And that explains the spectacular tail that sometimes can be seen when a comet comes close to the sun. When the comet is far away from the sun, the ice is frozen and the comet has no tail. When it comes close to the sun, the ice melts and evaporates. The vapor and some dust that it carries form an atmosphere around the comet. But comets are too small to hold the atmosphere by the force of gravity around them. So this atmosphere, consisting of gas, escapes into space. Light reflected from this gas provides us with a bright spectacle. The atmosphere of the comet escapes in a direction opposite to the sun. So the comet's tail always points away from the sun. This is because the sun bombards the comet with light and other particles (such as electrons) that push the atmosphere away from the sun.

What is the shape of the trajectory (or orbit) of a comet? It is an ellipse.

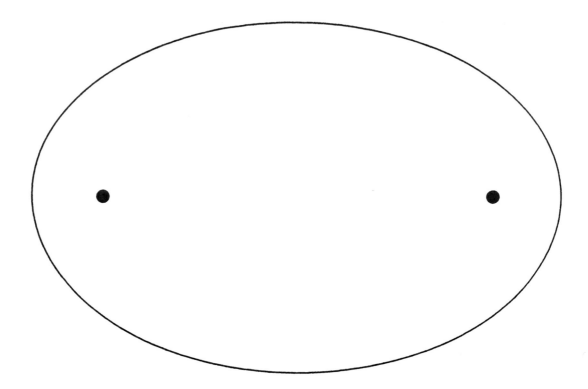

ILLUSTRATION 57. A curve that is an ellipse. The dots are the foci of the ellipse.

What Is an Ellipse?

Where do you get an ellipse? Here are three suggested answers:

- shadow of a circle
- oblique cross-section of a cylinder
- some cross-sections of a cone (other cross sections of cones are called parabola and hyperbola; together the three types of cross sections are called conic sections)

We will be concerned with a different description of an ellipse.

Take two points on the plane. And look at the curve consisting of all points such that the sum of the distances from those two points (the foci of the ellipse) is a constant. That curve is an ellipse (see Illustration 57).

Notice that as the foci get closer and closer together, the ellipse looks more and more like a circle. And if the foci coincide, the ellipse is a circle. So a circle is a special case of an ellipse, where both foci occupy the same point.

Drawing an Ellipse

This property of an ellipse gives a nice way to draw one (see Illustration 58). Fix a piece of paper to a table by putting tape on its corners. Cut a piece of string and fix its two ends very securely with transparent tape on the sheet. The loose string between the two pieces of tape is going to be used to create the constant distance, and the two points where the tape is holding the string to the paper are the foci. Make two dots with a pencil to mark the foci. Now put the loose string over the point of your pencil and put the pencil

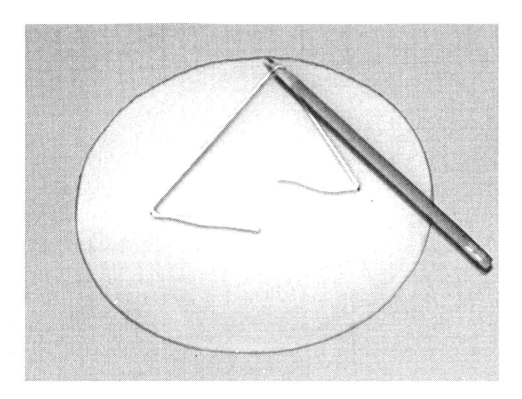

ILLUSTRATION 58. Drawing an ellipse.

on the sheet of paper, pulling the string taut. Move the pencil around, making a mark and keeping the string taut. Your finished figure should be an ellipse.

You might experiment with different distances between foci and different amounts of string. Skinny ellipses are harder to draw.

Kepler's Law and Interpreting What We Know about Comets on the Drawing

When Nicolaus Copernicus (1473–1543) put the sun in the center of the solar system, he assumed that planets move around the sun in circles. Johannes Kepler (1571–1630), looking at the data of Tycho de Brahe, and doing some shrewd guessing, postulated that orbits of all bodies moving around the sun are ellipses. The sun is always at a focus of an ellipse (the other focus is irrelevant). For planets close to the sun (such as Mercury or Earth), the ellipses are so close to a circle that it is really difficult to distinguish them from circles. But even with these scanty data, Kepler formulated his law, which tells how fast an object (a planet or comet) is moving around its ellipse. If the ellipse is just a circle, the object moves with a constant speed. But in the more general case, the planet (or comet) moves with such a speed that the radius from the planet to the sun sweeps out the same area for each unit of time (see Illustration 59).

This insight was theoretically explained by Isaac Newton's (1642–1727) famous law of gravity, which states that every two bodies in the universe attract each other with a force proportional to the product of their masses, and inversely proportional to the square of their distance from each other. When Newton said that it is easy to see far when you are standing on the shoulder of giants, he meant mainly Kepler.

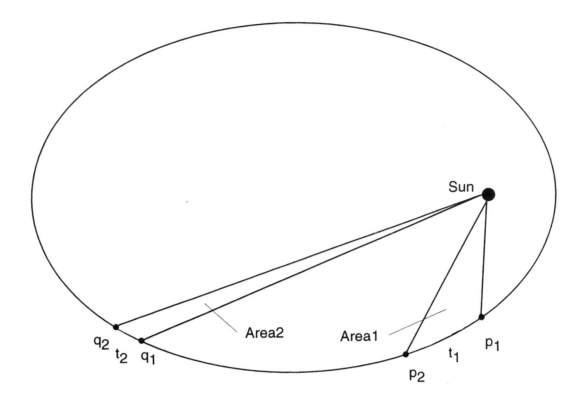

ILLUSTRATION 59. The time it takes the comet to go from p_1 to p_2 is the same as the time it takes to go from q_1 to q_2, so the areas 1 and 2 are equal.

Take an ellipse that you made, or make a new one. (Try to make it skinny.) Mark the sun as one of the foci. Now let's see about the comet when it is moving around its trajectory.

Activity 1

When does the tail develop, and in which direction does it point as the comet moves around the sun? Draw a line from the sun to the position of the comet. Draw a tail in the form of a plume. It is along a line starting from the comet and pointing away from the sun (see Illustration 60).

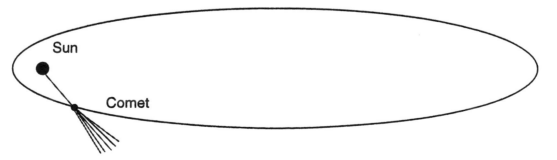

ILLUSTRATION 60. A drawing of the sun at the focus of an ellipse and a comet with a tail pointing away from the sun.

Notice that after the comet passes the point that is called the perihelion (its closest encounter to the sun), the tail starts "running the dog." (Some people think the tail is trailing a comet like dust behind a car. This is not so. Only the position of the comet relative to the sun decides the direction of the tail.)

Activity 2

Choose two points on the trajectory that are relatively close to each other, p_1 and p_2. And choose another two points, q_1 and q_2, also relatively close to each other but in a different part of the orbit (see Illustration 61). Compute how much longer (or shorter) it would take the comet to travel the distance from q_1 to q_2, compared to the distance from p_1 to p_2. When the points p_1 and p_2 or q_1 and q_2, are close enough, you need only to consider the area of two triangles. Each triangle has two points that are beginning and end points on the trajectory, and the third point is the sun. Use any technique that you know to measure and compare the areas of the two triangles. The ratio of the areas is, by Kepler's law, the ratio of time that it takes to pass through those segments of the trajectory.

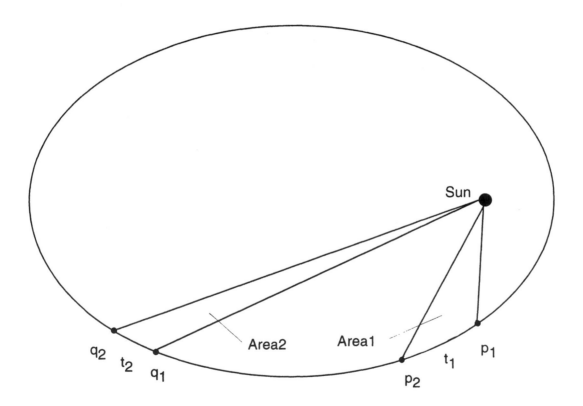

ILLUSTRATION 61. The ratio of the two areas, Area1/Area2, is the ratio of times that it takes to pass through those segments of the trajectory.

CONSTRUCTIONS

Two-Dimensional Constructions

CHAPTER 40

Circles

These activities may be started in kindergarten. The later ones are suitable for upper elementary and even middle school.

PROPS

Compass, ruler, and protractor (one per child).

Comment: Measuring angles and dividing a circle are intrinsically connected. Therefore, learning how to use a compass and a protractor at the same time has some merit.

THE LESSON

Exercises (see Illustration 62)

(1) Draw several circles of the same size.

(2) Given a circle, draw another of the same size.

(3) Draw a diameter of a circle.

(4) Draw two diameters and measure all four angles between them. (The circle should be slightly bigger than the protractor. This shows that you measure angles "around a point.")

(5) Draw two radii (half of a diameter) and measure the two angles between them (one of them has more than 180 degrees.)

(6) Draw three radii. Measure two adjacent angles. Add the values. Measure the union of the angles. (It should equal the sum you computed.)

(7) Draw a circle and several others with the same diameter, putting the point of the compass on the edge of the first circle. All these circles intersect in the center of the first one.

(8) After drawing a circle, use the compass to mark 6 points on the circumference (make any consecutive two be a radius apart). Connect them, getting a hexagon. Draw three diameters. Measure the angles of the triangles (60 degrees each).

(9) Draw a circle and a diameter. Using the diameter as a base, draw a triangle with the third point being somewhere on the edge of the circle. Measure all three angles of the triangle. One of them is a right angle. (Is this always the case? Yes.)

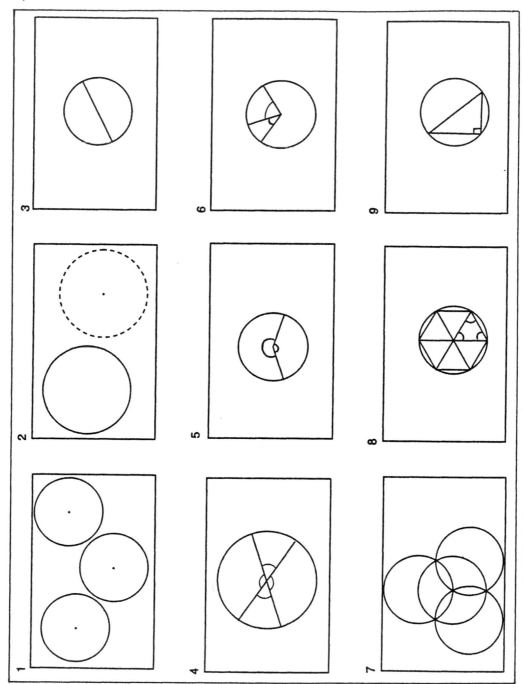

ILLUSTRATION 62. Diagrams 1 through 9 correspond to exercises 1 through 9 given in the lesson.

172

CHAPTER 41

Hanging Star

This lesson was taught in two second grade classrooms. It was taught in December, before the holiday break.

PROPS

Each child should have an 8 ½ by 11 inch sheet of light-weight white poster board, containing a drawing of a circle (approximately 19 cm or 7 ½ inches in diameter), with five marks equally spaced around its circumference (see handout at end of lesson). Each child also needs a pencil, a ruler with centimeters, a scissors, some crayons or markers, and a calculator.

THE LESSON

The teacher should have one or more finished stars, each cut from a piece of similar posterboard. (The posterboard cut away from each star will come in handy too; see below.) Each finished star should have yarn glued around its edge and a hole punched in the tip of one point. Through the hole should be threaded a piece of yarn, tied to form a loop for hanging (see Illustration 63). She also needs a transparency of the circle with marks, a transparent ruler, colored markers, scissors, a paper punch, two balls of yarn (red and green), glue sticks or other glue for the children to use, and an overhead calculator and projector.

I hung the finished stars on lines that were strung across the classroom and said, "Look what I have! Would you like to make one to take home?" The children were in enthusiastic agreement that they would like to do it. "Okay, we will need a few things to be able to do it. I've got some heavy sheets of paper here, with circles drawn on them, and some rulers, pencils, and scissors. Let's pass them out." After they were distributed, I said, "Do you think we'll be able to make a star with the things we have?"

"No," said several. "There is not a star on the sheet."

"Wait," said others. "There are some marks on the edge . . . "

"How many marks?" I asked.

"Five!" they replied.

"Can we make a star using them?" I asked.

"Maybe," said some.

"Hmmm," I said. "What do you notice about the star?"

"It has five points," said Brenda. "Maybe we can make a point at each mark."

173

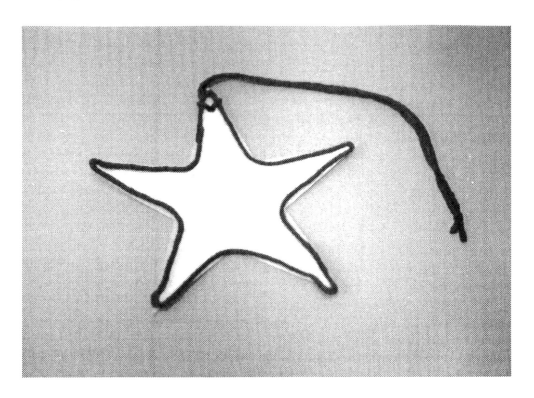

ILLUSTRATION 63. Hanging star.

"Great idea!" I said. "How shall we start? Let's see, if I draw a line with my ruler from every mark on the circle's edge to each mark in turn, will I get a star?"

"No," said Stan. "You need to make the lines cross."

"So what should I do, then?" I asked.

"Maybe you need to start at a mark, and then skip a mark; join the next one over," said Carrie.

"Let's try it," I said. "I'll do it on my star on the overhead, and you can watch if you are not sure how to do it. Do you know how to draw a straight line with a ruler?" I asked. "I bet some of you have never done it before."

"Pick a mark—any mark, and line your ruler up with it and with, not one right next to it, but one that is one mark away, like this," I demonstrated. "Hold the ruler very firmly. Then run your pencil close to the ruler, from one mark to the next, like this." I drew a line, and then began to help children as they drew. Many could not draw straight lines and needed some help—the main problem was holding the ruler so that it did not move. It was a good thing the pencils had erasers on them! Eventually everyone had one line drawn. "Okay, let's continue," I said. "Let's start at one end of the line we just drew, and make another, like this." I demonstrated, skipping an adjacent point and connecting with the second point over. "See how it looks like part of a capital A." Many children required help again, both in holding the ruler and in marking with their pencils along the edge of the ruler. But after a while all children had drawn ∧. "Next let's make the cross in the A. Do you see the two marks that run across?" I asked. I drew my line on the overhead and began to help the children.

"Hey," said Jeremy. "I think I see a star coming out!"

"Hooray," I said.

"What's next?" Jeremy said, as he pointed, "Look, there are only two more lines, from here to here, and here to here!"

I asked him to come up to the overhead and place my ruler correctly for one of the lines. He did so, and I drew. "Can you follow?" I asked the children. Most of the kids were able to do so, but some still required help.

"Now, can everyone see where the last line should go?" I drew the last line, and the star was complete. Again, most children could follow, but some required help.

"Cool," said several, "We have drawn a star!"

The teacher and I helped until each child had drawn a star. We used erasers a lot!

"Now I suggest you cut out the circle," I said. "It will make cutting out the star easier." We all cut out our circles. I then took five "pie-shaped" posterboard wedges, which I had cut from the stars I brought to class, and placed them on my circle on the overhead, so that the star shined through. "Look," I said. "Now we need to cut out these five pieces, in order to get our star,"

"Oh," said Annemarie. "I get it!" (This seemed to be a very important step in helping children see what to cut.) Pretty soon, everyone had a star.

"When you get your star cut out," I said, "would you like to color it?"

"Yes," came the unanimous reply.

We passed out colors, and I colored my star on the overhead. "You can decide which of the star's points is going to be the top of your star, and I will come around with a paper punch and punch a hole in it," I said. "Then I will give you a piece of yarn to put through the loop."

Coloring the star was a wonderful activity. Some children decided to color both sides. Many put their names on their stars. I walked around and punched holes, and the teacher and I passed out yarn to make a loop for hanging. Several children asked for enough yarn so that they could hang their stars around their necks!

Several said, "We want to put yarn around the border, too, so it will look like the ones you brought!"

"Okay," I said, "I will give you yarn, but here is what you have to do. You have to tell me how much yarn you need. I will then measure it, and cut a piece that is the right length. Then you can glue it on."

"Uh oh," said Brenda. "How do we figure out how much yarn we need?"

I took down the stars that were hanging. "Well," I said, "I guess we want yarn all along the border of the star—along all of its sides."

"Yes, but how long is that?" asked Brenda.

"How many sides does the star have?" I asked.

This question caused some consternation.

"It has five!" said several.

"No, it has five points. But it has more sides than points," said others. "Let's count the sides," I said. I put my star in the overhead, and we began to count: "One, two, . . . , nine, ten!" It has ten sides.

"How long is one side?" I asked. "Can you use your rulers to find out? Let's use the centimeter side of the ruler."

"What do I measure?" asked David.

I demonstrated on the overhead. "Pick any tip you want, and measure a side, from the tip." I laid my ruler on the overhead.

"It is about 7 centimeters," said Jeremy.

"Does everyone agree?" I asked. "Do you see the '7' on your ruler?" There was a pretty good consensus here. "Okay, let's say there are 7 centimeters on a side. How many sides?"

"Ten," replied the class.

"So how much yarn do we need? If you can figure it out without your calculator, do it. Or you may use your calculator."

"We can add up 7 plus 7 plus 7 ten times," said Carrie. "I know how to do that on the calculator: [7][+][=][=][=][=][=][=][=][=][=][=]."

"Why don't you try it, and see what you get?" I asked.

"Seventy!" she said.

"Seventy what?" I asked.

"Centimeters!" she said. "We need 70 centimeters of yarn."

"Great," I said. "I will show you a shortcut," I said. "Do you know about the 'times' key?" Some children did, but most did not. "Well, let's use it to compute how much yarn we need. We have 10 sides, and each side is 7 cm long. Try this: [10][*][7][=]."

"Seventy!" replied the children.

"Yes," I said.

The teacher and I measured out 70-cm lengths of yarn and went around to the children. "You can have your choice of red or green yarn," I said.

By this time, it was time for lunch. The teacher said, "Leave your stars and yarn on your desks, and after lunch we will paste the yarn onto the stars."

The lesson was a huge success. The children were not eager to go to lunch! They swung their stars to and fro and took great pride in what they had created.

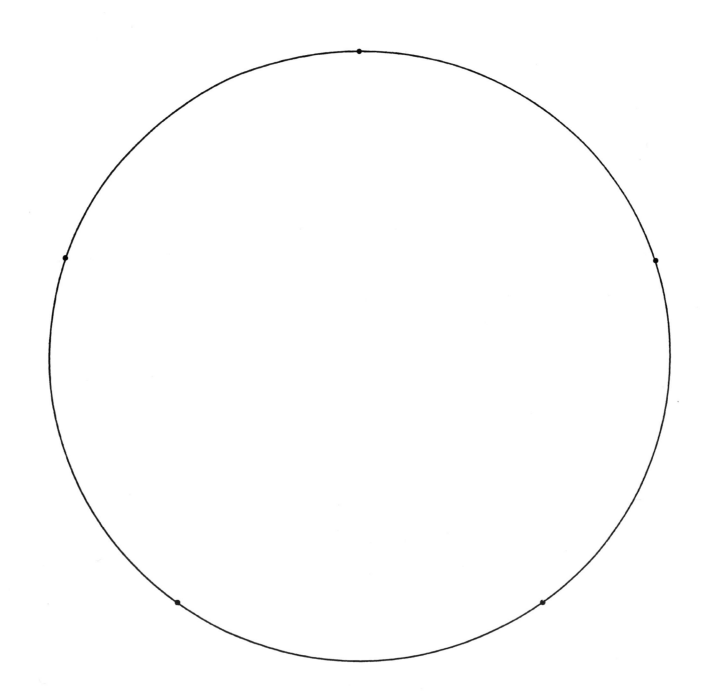

CHAPTER 42

Yin and Yang

This lesson has been taught in a fourth grade class. The activities near the end are suitable for middle school and higher.

PROPS

For each child: One or more sheets of blank typing paper, one sheet of thick white construction paper, a compass, ruler, crayons or markers, and scissors (see Illustration 64).

THE LESSON

Do you know the Chinese sign Yin and Yang? (See Illustration 65.) Study it. Here is something about it, from the *Random House Dictionary:* "Yin and Yang (in Chinese philosophy and religion), two principles, one negative, dark, and feminine (Yin) and one positive, bright, and masculine (Yang), whose interaction influences the destinies of creatures and things."

Would you like to make a yin and yang symbol? In order to do so, we need to be able to draw circles with a compass. Once we get a good large circle drawn on our thick paper, we can work on the inside. On the inside, the dividing line between the black and white parts consists of two half circles.

Activity 1: Learning to Draw a Circle with a Compass

Children should have one or more sheets of white typing paper. They should practice drawing circles until they get one that is large and essentially perfect. When they are able to draw a large circle, they should draw one on their piece of thick construction paper, and with a pencil lightly draw a diameter (a straight line through the center of the circle which touches both sides).

Activity 2: Drawing Two Half Circles Inside the Big Circle (see Illustrations 64 and 66)

Children should measure the diameter of their circle and put two marks on the diameter, 1/4 of the distance from each end. Computation can be done with a calculator. For example, if the diameter is 18.7 cm, the keystrokes are [18.7][/][4][=] (display: 4.675). So marks should be placed at about 4 cm 7 mm from each end of the diameter. These marks will be the centers of the two half circles.

177

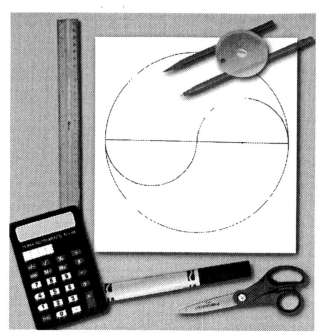

ILLUSTRATION 64. Drawing two half circles inside a bigger circle and the supplies necessary.

ILLUSTRATION 65. Yin and Yang.

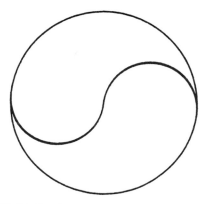

ILLUSTRATION 66. Drawing two half circles inside a bigger circle.

The diameter of each half circle should be 1/2 of the diameter of the original circle. One half circle should be drawn on one side of the diameter, and the other should be on the other side. Children should adjust their compasses so that the radius of the half circle is 1/4 of the diameter of the original circle. This can be done by putting the point of the compass on one of the marks, and the tip of the pencil on the circumference of the large circle (see Illustration 64).

Drawing the half circles requires precision, so erasers should be available.

Activity 3: Decorating the Drawing

Erase the diameter (see Illustration 66). Color half of the picture with a dark crayon or marker, and cut it out.

Activity 4: Dividing the Drawing

Now pretend that it is a cookie covered with white and dark chocolate. You want to share it with a friend. Figure out how to cut it into halves so that both of you will have the same amount of the white and the dark part. Can you cut it with a straight line? a curved line? When you figure out a way to do it, cut the cookie.

Remarks: We wrote the following words on the board during the lesson and gave their definitions:

- compass: a tool for drawing circles
- diameter: a straight line drawn through the center of a circle
- radius: one half of a diameter; the length of a line from the center of a circle to its edge
- circumference: the border of a circle

SOLUTIONS FOR ACTIVITY 4

For the curved lines (see Illustration 67), all four parts are congruent. (Make a big yin yang symbol, and cut along the curved line!) One half of a small circle has an area 1/8 of the whole big circle. Add to it another 1/8 (in the form of a section of the big circle),

ILLUSTRATION 67. A curved cut that divides the symbol into two halves with the same amount of white and dark.

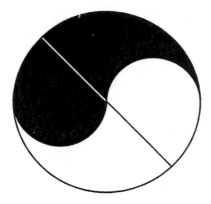

ILLUSTRATION 68. A straight line that divides the symbol into two halves with the same amount of white and dark.

and you have a fair share of dark (or white) chocolate, which is 1/4 of the whole big circle.

For the straight line (see Illustration 68), 1/2 of a small circle has an area 1/4 of one half of the big circle. Add to it another 1/4 (in the form of a section of the big circle) and you have a fair share of dark (or white) chocolate.

CHAPTER 43

Olympic Rings

This lesson has been taught in grades two and up.

PROPS

Each child should have five squares of posterboard, approximately 4 inches square, in each of the five colors blue, yellow, black, green, and red; good scissors, a ruler, and a compass. Cellophane tape. Yarn for tying the rings into a necklace (optional). A sheet of white paper to glue the finished rings on (optional).

THE LESSON

The summer Olympic Games were held in Atlanta in July 1996, and will be held in Sidney, Australia, in the year 2000. You have probably seen pictures of the five rings that appear on the Olympic flag. The flag was adopted in 1913. The rings were selected to symbolize the unity of the five continents coming together in peace to celebrate athletic excellence. The colors of the rings: blue, yellow, black, green, and red, were selected because at least one of those five colors occurred in each of the participating nations' flags.

A picture of the rings is shown in Illustration 69. Do you see how they are interlocked to make a chain?

We are going to make five colored rings out of posterboard, and interlock them in the same way.

Part 1: Drawing and Cutting out the Rings

We need to find the center of each square piece of posterboard; we'll put the point of our compass on this spot. One way to find the center is to draw lightly two diagonals; where they intersect is the square's center.

Draw a circle with as large a radius as you can on each of the five squares. The radius will be between $1 \frac{3}{4}$ and 2 inches. Draw a second smaller concentric circle (a circle having the same center as the larger one) on each square, with a radius of about 3/16 inch less. Be sure that the circles on all five of the squares are identical in size. Now cut out the rings. Cut carefully! This takes some patience, as the posterboard is difficult to cut.

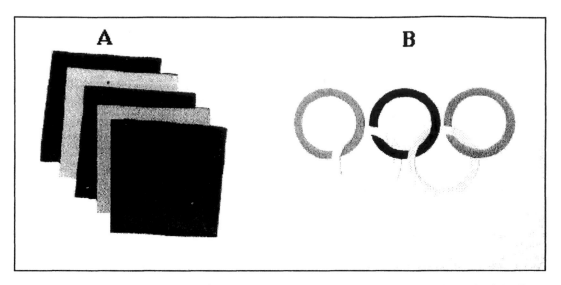

ILLUSTRATION 69. (A) Five squares of poster board, approximately 4 inches square, in each of the five colors blue, yellow, black, green, and red. (B) Olympic rings.

Part 2: Interlocking the Rings

See Illustration 69. The colors of the rings, from left to right, are blue, yellow, black, green, and red. You need only to cut two rings (the yellow and green ones) in order to make a chain. The blue ring goes over the yellow, yellow over black, black over green, and green over red. Tape the cuts. If desired, you may tie some yarn on two of the rings, in order to make a necklace. Which two rings do you choose? You might also glue the rings together, to hold them in the position shown above, or glue them on a white sheet of paper.

CHAPTER 44

Snowflake

THE LESSON

Real snowflakes come in all shapes and sizes. There is a saying that no two snowflakes are alike. But usually during one snowstorm the snowflakes have a similar general appearance. Sometimes they are small and powdery like sugar. Sometimes they look like small grains of cereal. Sometimes they are large and flat and fall very slowly. If we look carefully at flat snowflakes (we need a magnifying glass if they are small) we see that their overall shape is usually hexagonal (see Illustration 70) and not square, triangular, pentagonal, or any other shape.

Task 1: Free-hand drawings

Children should try a free-hand drawing of an equilateral triangle, a square, a regular pentagon, and a regular hexagon.

Task 2: Drawing a regular hexagon with a compass and ruler

Draw a circle. Mark six points along the circle, any consecutive two being a distance of one radius apart (see the picture in Illustration 71 of a circle with points marked). Connect the points using a ruler to make the sides straight.

Task 3: Getting familiar with a hexagon

Each child will need a compass, ruler, scissors, an envelope, and colored paper.

Each child should draw and cut out four hexagons of different colors, 2 to 3 inches across. One should be cut into two congruent parts (along a diagonal). The second one should be cut into three rhombuses (squished squares). The third should be cut into six equilateral triangles. The last should be cut into twelve small right triangles (see Illustration 71, bottom).

Each child should reassemble the hexagons again (as puzzles), and then put them into the envelope to take home.

Task 4: Making a paper snowflake

Each child will need the same items as in Task 3 and one large piece of white paper.

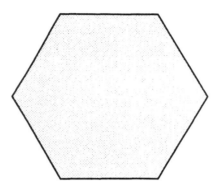

ILLUSTRATION 70. A hexagon.

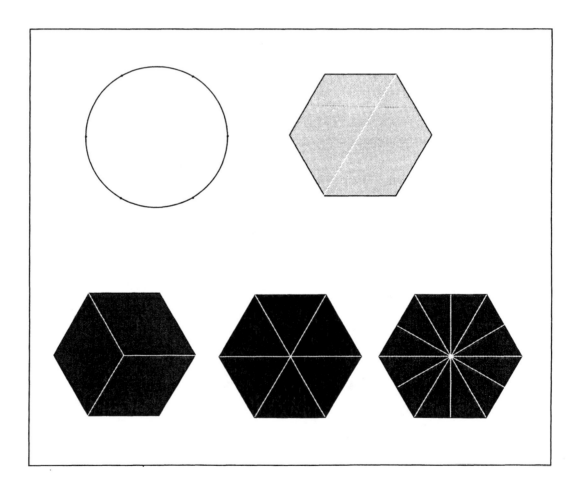

ILLUSTRATION 71. Upper left: a circle with six points marked, any consecutive two being a distance of one radius apart; upper right: a hexagon divided into two trapezoids; bottom left: a hexagon divided into three rhombuses (it looks like a cube); bottom middle: a hexagon divided into six congruent equilateral triangles; bottom right: a hexagon divided into twelve congruent right triangles.

PART 1

Draw and cut out a large hexagon. Fold it into a triangle (first in half, then into three parts) (see illustration 72, which shows the stages of folding).

PART 2

Cut some patterns from the edges of the folded triangle. Unfold it, and you have a giant paper snowflake (see Illustration 73). Are any two snowflakes exactly the same?

Remark: This task is more difficult than it seems. The hexagon must be regular; the folding must be precise; and good quality scissors are important.

With free-hand cutting of the edges of the folded triangle, it is easy to to cut the triangle into separate pieces, and to destroy the snowflake. Some children may have to try more than once to succeed, and some may need help. The teacher should have some hexagons ready, so if a child messes up the cutting, he or she can get another hexagon from the teacher, and will not have to start from the beginning.

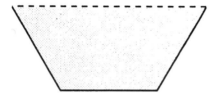

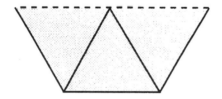

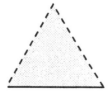

ILLUSTRATION 72. Three steps in folding a paper snowflake. (1) Fold the hexagon in half; (2) find the midpoint of the big side of the trapezoid and make creases to form an equilateral triangle; (3) fold into an equilateral triangle.

ILLUSTRATION 73. Some paper snowflakes.

186

CHAPTER 45

Fractions of a Circle, AKA Caterpillar

PROPS

Brightly colored poster board (white on one side), cut into squares 5 by 5 inches. Four colors are needed (for example, yellow, green, orange and pink). Envelopes. Compasses, protractors, rulers, scissors (good quality), calculators.

THE LESSON

Task

Children work in pairs. Each pair gets four squares of different colors, two envelopes, and a set of tools. The children are asked to draw and cut out four circles, each four inches in diameter. The circles have to be divided into sections as follows:

- Yellow—four quarters (1/4 of a circle each)
- Pink—eight equal parts (1/8 of a circle each)
- Orange—eight unequal parts (four parts, 1/6 of a circle each; four parts 1/12 of a circle each)
- Green—eight unequal parts (two parts 1/5 of a circle each, six parts 1/10 of a circle each)

Each child gets half of all the pieces and keeps them in his or her envelope.

Method

Compute the appropriate angles. Draw the circles on the white side of the squares. Using a protractor, draw the angles within the circles. Carefully cut out the circles and cut them into sections (see Illustration 74).

Part of a circle	1/4	1/8	1/6	1/12	1/5	1/10
Angle in degrees:	90°	45°	60°	30°	72°	36°

Remark. If children are unskillful with tools, they may start by drawing the plans on normal white paper before copying them onto poster board.

Activities with Sections

Order the sections according to the sizes of the angles. Make two whole circles in many different ways (see Illustration 75). Construct new angles given their measure

187

ILLUSTRATION 74. Four circles cut into sections.

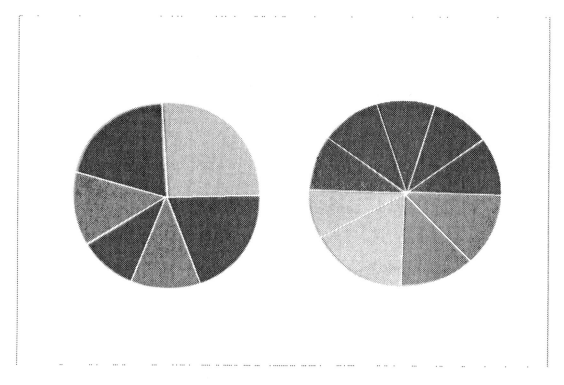

ILLUSTRATION 75. Two circles made in two different ways.

188

ILLUSTRATION 76. A caterpillar made from sections of circles.

(examples: 147 = 45 + 30 + 72; 136 cannot be constructed). Compute the areas of all sections either in square inches or in square centimeters. Make interesting patterns (see Illustration 76 showing a caterpillar.)

Some Theory

This lesson deals mainly with skills with tools, but some comments about fractions are in place. In schools the word fraction is used in three different meanings:

(1) A fraction is part of an object. "Shade 1/3 of this circle." "Fractions of a circle."
(2) A fraction is a mathematical expression. "Write 0.2 as a common fraction."
(3) A fraction is a number. "Add 1/2 and 1/4." (You add numbers; the result is 0.75.)

Usually the meaning is clear from the context, but children can get confused and not know if we are talking about numbers or objects. We compute an angle (number) of 1/4 of a circle (object) by multiplying 360 (number) by 1/4 (number) or by dividing 360 by 4 (both numbers). Notice that the word angle is also used in three different meanings: a number, two rays, or the part of a plane between two rays.

CHAPTER 46

Square Inch Circles

INTRODUCTION

Many children and adults think that we can measure in square inches only those figures that can be exactly filled with 1 inch by 1 inch squares. In this lesson children construct circles of required areas, and they also look at different approximations of π (pi). The lesson is suitable for grades six through eight.

TASK

Construct 5 circles, having areas 1, 2, 3, 4, and 5 square inches (see Illustration 77).

PROPS

Colored construction paper, envelopes, compasses, rulers, and calculators.

THE LESSON

Method

- Children compute radii of circles with an accuracy of 1/16 of an inch, using three different approximations of π, 3.1415926, 3.14, and 3.
- They draw and cut out circles from construction paper and write on each one its area (e.g., 2 sq. in.).
- They write their name on an envelope and put the circles inside.

Computation of Radii

General formula for the radius r of a circle with area A,

$$r = \sqrt{(A/\pi)}$$

191

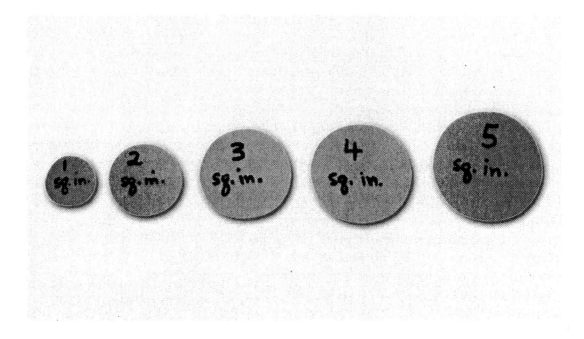

ILLUSTRATION 77. Five circles, having areas 1, 2, 3, 4, and 5 square inches.

A program for computing r in inches:

Keystrokes	Comments
[MRC][MRC]	
[approximation of π][M+]	Put an approximation of π in memory;
For each value of A,	
[A][/][MRC][=][√]	compute the radius r, for the area A, name the whole part of the answer w;
if $w > 0$,	
[−][w][*][16][=]	round the answer to the nearest whole number,
else,	name it p; this gives you the fraction part of
[*][16][=]	the answer.

The answer is: The radius $r = w + p/16$ inches.

Example 1

Computation for $\pi = 3.1415926$ and $A = 2, 5$.

Keystrokes	Display	Comments
[MRC][MRC]		
[3.1415926][M+]	3.1415926	
[2][/][MRC][=][√]	0.7978845	$A = 2, w = 0$

Keystrokes	Display	Comments
[*][16][=]	12.766152	$p = 13$, $r = 13/16$ in.
[5][/][MRC][=][√]	1.2615662	$A = 5$, $w = 1$
[−][1][*][16][=]	4.1850592	$p = 4$, $r = 1\ {}^4\!/_{16} = 1\ {}^1\!/_4$ in.

Example 2

Computation for $\pi = 3$ and $A = 4, 5$.

Keystrokes	Display	Comments
[MRC][MRC][3][M+]	3	
[4][/][MRC][=][√]	1.1547005	$A = 4$, $w = 1$
[−][1][*][16][=]	2.475208	$p = 2$, $r = 1\ {}^2\!/_{16} = 1\ {}^1\!/_8$ in.
[5][/][MRC][=][√]	1.2909944	$A = 5$, $w = 1$
[−][1][*][16][=]	4.6559104	$p = 5$, $r = 1\ {}^5\!/_{16}$ in.

CHAPTER 47

Fractions with Popsicle Sticks

This lesson has been taught in fourth and fifth grades. It takes a full hour, and could be spread over two days.

ORGANIZATION AND PROPS

Children, work in groups of about 4. Each group has a set of 8 colored markers. Each child has a calculator, a ruler with centimeters and millimeters, and 8 popsicle sticks. The teacher should also have his/her own collection of colored popsicle sticks, marked with halves, thirds, fourths, fifths, sixths, sevenths, and eighths, and nicely colored (see Illustration 78), as explained below. (Children are given 8 popsicle sticks rather than 7, so that if they make an error, they will have a spare. It is a good idea to have even more spares available! A box of 1000 popsicle sticks can be bought for about $3 at craft stores.) An overhead projector and overhead calculator are also used, together with a transparent ruler for the overhead.

THE LESSON

Part 1: Fractions with Popsicle Sticks

I took my 7 popsicle sticks from an envelope and showed them to the children. (See the illustration.) "Look what I have here," I explained. "I have a collection of 7 popsicle sticks. This first one is divided into two equal parts, and one half of it is colored blue. I've written 'halves' on the back. The next one is divided into three equal parts, and I've colored every other section, and written 'thirds' on the back. The next is in four equal parts, and I've colored every other section. What do you think I have written on the back?"

"Fourths!" answered the class.

"And the next?"

"Fifths!"

"And the last three?"

"Sixths, sevenths, and eighths," they answered.

"Would you like to make a collection for yourself, and color them?"

"Yes!" they responded.

"OK. Let's start with halves." I held up my stick with halves. "What do we need to do?"

"Well," said Betsy. "We need to find the middle of the stick."

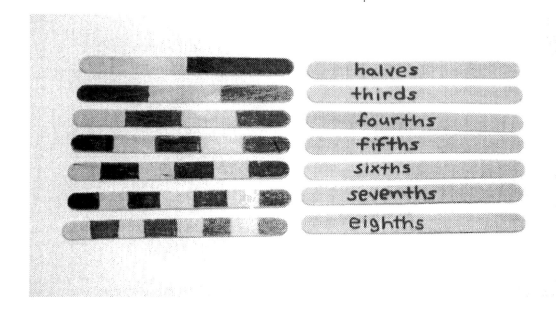

ILLUSTRATION 78. Popsicle sticks divided into halves, thirds, fourths, fifths, sixths, sevenths, and eighths.

"How do we do it?" I asked.

Nakia said, "We find out how long the stick is, and we divide it by 2!"

"Great!" I said. "How do we find out how long the stick is?"

"We measure it!" said several.

"OK, how long is it? Let's use the centimeter side of our rulers." The children measured and decided a stick was a little longer than 11 centimeters. I drew on the board:

$$| \ldots | \ldots |$$
$$11 \qquad 12$$

"Do you see the little spaces here, between 11 and 12?"

"Yes, they are millimeters," said several.

"How many are there between 11 and 12?"

"Ten."

"And where does the end of the popsicle stick land?"

"About two or three millimeters past 11," they said.

We voted and decided we would say the popsicle sticks were 11 centimeters and 3 millimeters long. "OK, how do we write that as a decimal?" I asked.

"11.3," they answered.

"Now we want to find the middle of the stick. How do we do it?"

Several knew the calculator keystrokes: [11.3][/][2][=]. I wrote it on the board. The display showed 5.65. I wrote it on the board. "What does that mean?" I asked.

Lance replied, "On your stick, you make a mark at 5 centimeters and half-way past six millimeters."

"Does everybody get that?" I asked. I showed it on the overhead with my transparent ruler, and the teacher and I went around to desks to help children find their marks. "When you have your mark, you can color your stick," I said. "And I wrote 'halves' on the back of mine. You can too." I wrote *halves* on the board.

"My next stick is divided into thirds," I said. "How do you think I do it?"

"You put 11.3 into your calculator, and divide it by 3," said several.

"That is a good start," I said. On the board I changed the 2 to a 3: [11.3][/][3][=] display: 3.7666666. I wrote it on the board. "What is that?"

"It is 3 centimeters and 7, or maybe 8, millimeters!" answered several.

"OK, let's put a mark on our stick at 3 point 7, or better yet, 3 point 8, centimeters. Are you with me?"

"No," said some.

"Yes," said others.

After all had one third marked, I asked what we had to do next. "We have to find the 2/3 mark," said one. "How do we do it?" I asked.

"Well, we have one third," said Cora. "So we should just multiply it by 2 to get two thirds."

"Let's try it," I said. I wrote [*][2][=]. The display shows 7.5333332, which I wrote on the board. "That is about 7 and a half centimeters," said several. The children measured and made their marks, and then colored their sticks. "Let's write 'thirds' on the back," I said. I wrote *thirds* on the board.

"What is next?" I asked.

"Fourths."

"Who can tell me what to do?" They knew that I needed to modify the program on the board: I should change the 3 in the top line to a 4. So it looked like [11.3][/][4][=]. The display shows 2.825, which I wrote on the board. By this time most children were finding the marks fairly accurately and without help. "OK, after you find your one fourth mark, I want to show you something really sneaky. On your calculator display, do you have 2.825?" "Yes," they said. "Now press [M+]. Do you know what that does?"

"It puts 2.825 in memory, and causes an M to show on the display," said Gregory.

"That is right! And it will come in handy in a minute," I said.

"How do we get the mark for two fourths?" I asked. They knew the button presses were [*][2][=]. The display shows 5.65, and I wrote it on the board.

One child noted, "Hey, that's the same as we got for one half, on our first stick."

"Great observation," I said. "We'll come back to that."

The children knew they needed to find the 3/4 mark now, and they were unsure what to do. I said, "Try this: [MRC][*][3][=]. The display showed 8.475. The board now looked like this:

[11.3][/][4][=][M+]	2.825
[*][2][=]	5.65
[MRC][*][3][=]	8.475

I also wrote *fourths*. "Remember to clear memory before going on," I said. "Press [MRC][MRC]."

For fifths, I asked the children to tell me what to do. They modified the program so that it looked as follows (I wrote displays on the board):

[11.3][/][5][=][M+]	2.26
[*][2][=]	4.52
[MRC][*][3][=]	6.78
[MRC][*][4][=]	9.04

Interpreting 9.04 was tricky. Katherine said it was 9 centimeters and 4 millimeters. But Jim said no, it was just a tiny bit over 9 centimeters. One child said, it is like nine dollars and 4 cents; that is just about nine dollars. I also wrote *fifths,* and reminded them to clear memory before going on to sixths.

For sixths, the program looked as follows:

[11.3][/][6][=][M+]	1.8833333
[*][2][=]	3.76666666
[MRC][*][3][=]	5.64999999
[MRC][*][4][=]	7.5333332
[MRC][*][5][=]	9.4166665

By this time the children were measuring and marking with very little error, and they were becoming quite artistic in coloring their sticks. It was time for recess, so we decided to stop with sixths. But before quitting I asked them to put their halves stick next to their fourths stick. Several spontaneously observed that there are two fourths in a half. I also asked them to put their thirds sticks next to their sixths sticks, and again several noted that there are two sixths in a third.

Part 2: Other Programs for Finding Where to Put Marks

There are many programs children could use to get numbers telling them where to put marks on their sticks. For example, [1][/][2][*][11.3][=] will give 5.65.

Another program for finding fourths is as follows:

[11.3][/][4][*][1][=]	2.825
[2][=]	5.65
[3][=]	8.475

In one class a child noted that, for example, once she had the length for one third, she could just move her ruler up the popsicle stick, putting its zero at the previous mark, and measure another one third.

CHAPTER 48

Valentine Heart

This lesson has been taught in two second grade classes. It was taught on February 13, the day before Valentine's day.

Children draw a (stylized) heart, cut it out, trace around it to make a second heart, write messages on the two hearts, and calculate the area of each heart in square inches.

PROPS

Each child needs a pair of scissors, a ruler, a pencil, one piece of red paper containing a one inch grid (dots and not lines; inch dot paper is in the back of the book), and a one-inch square piece of paper in a contrasting color (we used white). We also had available one compass for each pair of children, which we handed out late in the lesson (see below). The lesson can be done without compasses if they are not available.

Before the lesson started, the instructor drew on the blackboard a 6 by 6 grid of dots. She used a meter stick to make the dots 5 inches apart. She prepared a piece of pink paper 5 × 5 inches square. She also hung picture 1 of Illustration 79 on the board using a magnet. And she prepared a "compass" by tying a string around a piece of chalk. Also she prepared an accurately cut out heart, which she cut into several pieces, as shown in picture 2 of Illustration 79.

THE LESSON

"Do you know what tomorrow is?"

"Valentine's day!"

"Today we are going to make some hearts out of red paper, and you can write messages on them for your boyfriend or girlfriend or mom or dad or anybody else you want. Does everybody have a ruler, a pair of scissors, a pencil, a sheet of red paper with dots on it, and a square inch? We will pass out calculators later if we need them. But I think today you can do the math that we will need just by using your heads."

"See this heart?" (She pointed to picture 1, tacked on the board.) "We are going to make it, and we will do it step-by-step. Here is the first thing you need to do. Make a square that is four inches long and four inches high. Watch me and I will show you how to do it." She used the grid on the board. "I will start up here in the corner and put my chalk on a dot. I count across: one step, two steps, three steps, four steps." She drew as she counted. "Now I go down: one step, two steps, three steps, four steps. Now I go back over, one step two steps, three steps, four steps. Finally, I go back to my starting place:

199

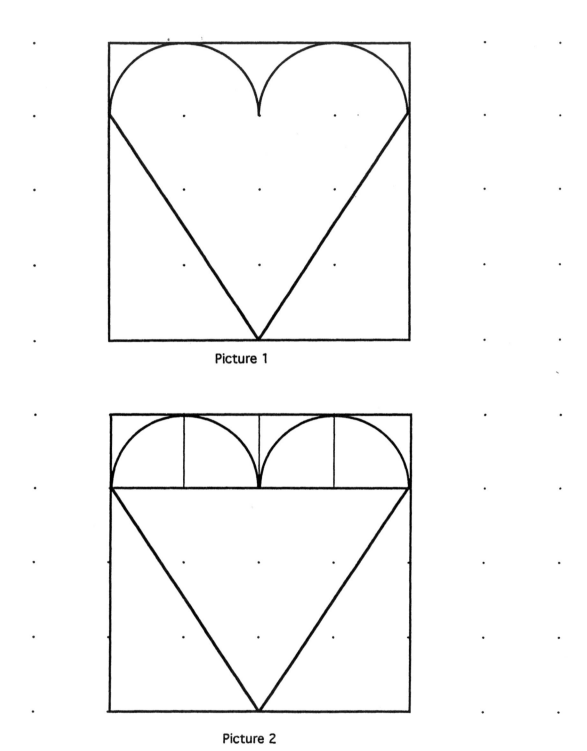

Picture 1

Picture 2

ILLUSTRATION 79. Picture 1 is a stylized heart. Picture 2 is a heart ready to be cut.

one step, two steps, three steps, four steps. Now you try it. Be sure you start on a dot, and start in a corner, because we need to save part of the paper for use later. You can use a ruler if you want."

The teacher and the instructor walked around to see that children were making the 4 × 4 inch square. Some children counted dots rather than inches, and ended up with a 3 × 3 inch square. The teacher said, "Let's talk about this. To get a line that is four inches long, you need to connect five dots, not four! Why? Because you need to take four steps, starting at the first dot.

"You can also check with your ruler. To make the top of the square, we want to draw a line that is four inches long through the dots at the top of our paper. Put the end of your ruler on a dot, and see where the 4 is: it is four inches away."

We waited until every child got a correctly drawn square.

"Now let's name this square. We will name it by its area, by the number of square inches that we can put in it. Take your square inch and count to find its name." The instructor took her "square inch" (actually five inches square) and counted, using the square she had drawn on the board: "One square inch, two square inches, . . . What is the total?"

The class responded, "16."

"Ok, so we will name our square sixteen. Its area is 16 square inches."

"Now we want to cut out our square." She pretended to "cut out" the square on the board, using a pair of scissors. "Try to cut straight along the sides."

"Now we are going to start to draw our heart," she said. (Illustration 80, a through f, shows the stages in drawing.) "First we are going to draw a line along the top of the square, just one inch down." Using a ruler, she drew on the board a line, as shown in Illustration 80b. "You can use a ruler to do it."

We waited until each child had correctly drawn the line.

"Now we are going to work on the bottom part of our heart." She pointed to the heart drawing. "We want to draw this line first." She pointed to the left line in Illustration 80c. "This is kind of tricky. Let's make an x right here (where the horizontal line they had just drawn hit the side of the square) and right here (on the bottom in the middle). Now you need to use a ruler to draw a straight line between the two x's."

This was quite difficult for children. They realized it was not a matter of simply connecting the dots: this particular line did not "go through" any dots in the interior of the square. Also, some children had never used a ruler before, and they needed help. For example, one child was able to hold her ruler so that a line could be drawn along it to connect the two x's. But when she actually drew the line, she did not put her pencil lead next to the ruler. The instructor helped her erase and draw the line again next to the ruler. Another problem was that, when the children cut out their squares above, sometimes they cut them a little small, so that the dots around the edge were no longer present. This presented a problem in finding the middle dot along the bottom. The teacher suggested, if the children could not find the middle dot, that they could fold the square in half and make a little pinch to indicate the middle.

"Now let's draw the line on the other side, to finish the bottom of the heart. Put an x on the side of the square, right here." She placed an x on the right side of the horizontal line. "Now draw a straight line between these two x's." She drew on the board.

When children had finished, she asked, "What shape have you drawn inside your square?"

They knew the shape and responded, "A triangle!"

"Now comes the hard part. We have to draw the curvy top part of the heart. Watch

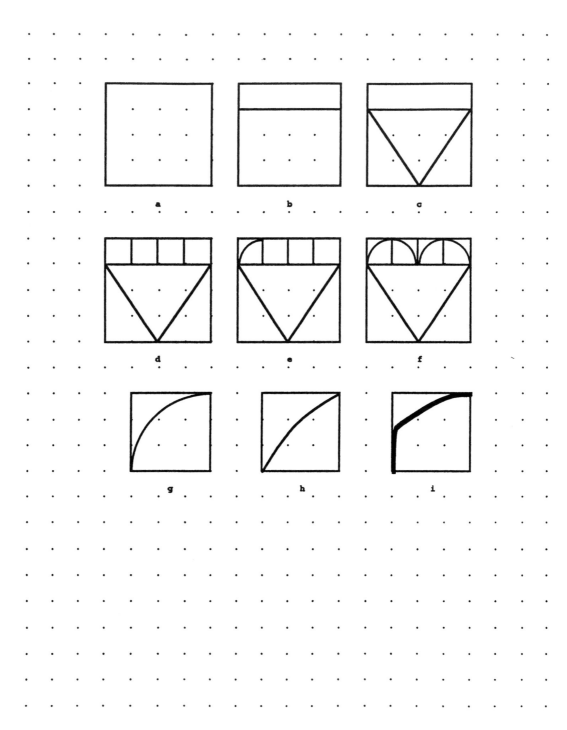

ILLUSTRATION 80. (a–f) The stages in drawing a heart. (g–i) Three arcs: (g) drawn correctly, (h, i) drawn incorrectly.

202

what I am going to do." Using a ruler, she drew the three vertical lines of length one inch, as shown in Illustration 80d. "I am drawing three lines. Each is one inch long. Try it. Use your ruler if you want."

Again we waited until all children had the correct drawing.

"How are we going to draw the curvy part? I am going to use a compass to draw a part of a circle. Have you ever used a compass?" A few children said yes, but most said no. The instructor said, "I have two compasses to show you. First let me show you how to draw a whole circle, using a string and chalk." She drew a big circle on the board. There were oohs and aahs. She then held up a student compass. She took a piece of white paper and held it against the board, and showed the children how to draw a circle using it.

"Now let me show you how to draw some of the curvy part of the heart using a compass." She used the chalk-and-string compass and drew a quarter of a circle, in the upper left square inch of her board drawing (see Illustration 80e). "Would you like to try it? We will pass out compasses now. You can use them, or you can draw your curve by hand. If you draw it by hand, be sure to make the curve nice and fat, as I have, and not squashed." We handed out compasses.

Getting compasses caused big excitement in the classroom. The children wanted to draw circles. "Ok, if you want to draw some circles, take the big part of your red paper and turn it over, and just draw some." They needed lots of help. Sometimes the teacher or instructor would need to hold the paper still or help the children adjust the pencil in the compass so that both it and the point of the compass could touch the paper at the same time. The instructor and teacher permitted children to draw circles for quite a while.

"Now you need to draw the first curvy part of the heart. Let me show you again how to do it." This time the instructor took a student compass, put its point on the correct dot on her drawing on the board, and swung an arc. "See what I have to do. I have to adjust the distance between my pencil and my point so that it goes from the dot to the the edge of my square. Try it. We will come around to help if you need it." The precision required here was very great. The problem was setting the radius of the circle, so that it would be the correct length (one inch).

"Now let's finish the curvy part. I need to draw 3 more curves just as before. Watch and I will show you how." The instructor drew the three remaining arcs. "See how I have to change where I put the point on my compass. Try it, and let us know if you need help." (See Illustrations 80f, 80g, 80h, and 80i; 80g shows a correctly drawn arc, while 80h and 80i are incorrect.)

Most children could do it, but they did not do it precisely. Their compass would slip, or their lines would not be continuous. Some children made rather large holes in their paper with the compass points. Some also moved the center of their quarter circles each time, making four copies of the quarter circle shown in Illustration 80e. The teacher and instructor helped until everyone had four correctly positioned "arcs" drawn.

"Ok, now we are ready to cut out our hearts. Be careful going around the curves!"

The children cut out their hearts.

"Now we are going to make another heart, but this time it will be quicker. First let's draw another square that is four inches tall and four inches wide. Do you still have room on your red paper to do it?" Some children did not, and they were given another piece of paper with dots. However, this time drawing the 4 × 4 inch square was easy. The instructor drew a 4 × 4 inch square (to scale) on the paper tacked on the board.

"Now we are going to put the heart we just cut out inside the square and trace around it." She demonstrated, using the picture taped to the board and a heart she had cut out beforehand.

This activity went fairly quickly. The children did not need help in tracing. "Now cut out your second heart." Every child in the classroom had two hearts. "Now you can write a message on each of your hearts. Decide whom you want to give them to, and write something!"

"One last question. Do you remember what the area of our square was? Do you remember what we named it? How many square inches does it contain?" They remembered its name was 16. "This is a hard question: What name can we give the heart? How many square inches do you think it contains? One thing we know: it is less than 16."

The children called out numbers: "6, 10, . . ."

"How can we find out? Let's start with the easy part first: the bottom triangle. Let me show you what I can do." She showed how the bottom triangle could be cut in half to form two congruent triangles, and that these two triangles formed a 2 × 3 rectangle. This caused some excitement. "What is the area of the triangle? What shall we name it?" The children were able to determine that it was six. [They had had the lesson titled Rectangles (Book I, Chapter 15) two weeks earlier.] "OK, the only thing left is how to name the curvy part. It is drawn in how many square inches? Let's count across the top: one, two, three, four. But the square inches are not all filled up. We are going to estimate here. About how much of the four square inches do you think are filled up with the heart?" Most children said about 3. "So what shall we name our heart? How much is six plus three? No calculators needed!" They knew the answer was nine. "The area of the heart is about nine square inches. Now let me show you just one last thing. Look what I can do with the four curvy parts." Using cut-outs, she showed how the four quarter circles form a whole circle.

"Be sure to deliver your hearts to the people you decided on!"

Remarks

This lesson was very successful in both classes. The actual area of the 4 curvy parts of the heart is about 3.1416 square inches (area = pi times the radius squared; here the radius is one, so the area is pi). Our estimate of three is very close. It is important to say that we are finding the areas (first of the square, and then of the triangle and the heart), and that we are naming the figures using the areas as the names.

CHAPTER 49

Cutting Simple Jigsaw Puzzles

This activity has been used in the middle elementary grades. It is also a good activity for Family Math Night.

THE LESSON

There are many situations in which precision in using scissors is an asset. In the following set of problems, acquiring manual skills is as important as is their intellectual content.

Problem 1

Each child gets 5 squares of paper of different colors, not necessarily the same size (the smallest commercially available origami paper is the best because its sides are two different colors; we have also used square sheets from memo pads), scissors, and an envelope.

TASK

Cut each square into four congruent parts (parts that have the same size and shape). But each square has to be cut differently! For some examples see Illustration 81. If the cuts are with straight lines, the paper should be folded and creased, and then cut along the fold. For more complex cuts, lines should be drawn first on the white side of the paper (if origami paper is used). Parts that are cut should be kept in an envelope. After the task is finished, children should "put together" the squares (as simple jigsaw puzzles).

Problem 2

Each child gets one large square of colored paper, scissors, an envelope, and two paper plates.

TASK

Cut the square into two parts of equal size (area), but not necessarily of the same shape; cut each piece into two parts of equal size; and again cut each new piece into two parts of equal size; and again cut each new piece into two parts of equal size. Put the pieces together to form a square. (At the end put them into the envelope and take them home.)

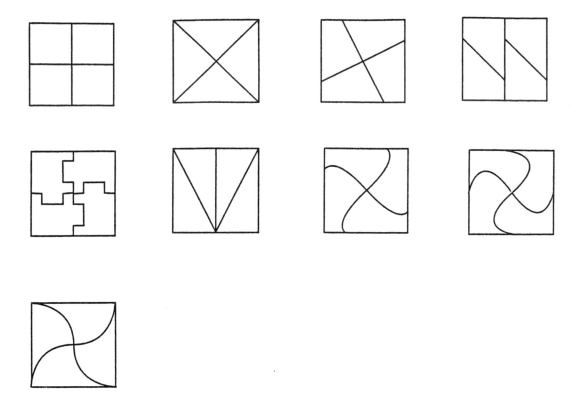

ILLUSTRATION 81. Squares cut into four congruent parts. Can you think of others?

SUGGESTED METHOD

It is easy to lose track of which pieces to cut. So after cutting the square, put both pieces on one plate. When you cut them, put the parts on another plate. Iterate this process so the next "generation" of pieces always lands on the other plate.

QUESTIONS

(1) How many pieces will you have?

Cut the square into 2 pieces.	[2]	display: 2 pieces
Cut each piece into 2 parts.	[*][=]	display: 4 pieces
And again.	[=]	display: 8 pieces
And again.	[=]	display: 16 pieces

Remark: There are two approaches. One (given above) is to see that by cutting, we double the number of pieces, and to use the calculator to get the exact number. The other one is to do the task and count the pieces. Both should be done. Doing the calculation before the task has the advantage of showing that it correctly predicts the outcome.

(2) Can you always cut a piece into two parts of the same size (area)? Yes. (Sometimes it is not so easy to find the right cut.)

(3) Can you always cut a piece into two congruent parts (of the same size and shape)? No. (Show such a shape.)

(4) Are all the small pieces of the same size? Yes. (Only approximately, because of the lack of precision in cutting.)

Remark: Do not be surprised if this question is difficult for children. There are two rather subtle principles involved here:

- the mathematical one: If $a = b$ then $a/n = b/n$.
- the physical one: If you cut a piece of paper, then the area of the original piece is equal to the sum of the areas of its parts.

Three-Dimensional Constructions

CHAPTER 50

Making a Cube

This lesson has been taught in fourth and sixth grade.

PROPS

Children had several sheets of one-inch square grid paper (see handout at end of book), scissors, pencils, and cellophane tape.

THE LESSON

A simple task was posed: "What patterns that we can draw on our dot paper and cut out will fold into cubic inches? How many patterns for cubes are there? When you think you have one, draw it on your paper, and then draw it on the blackboard. Then go to your desk and cut it out and fold it, to see if it really folds into a cube."

Children quickly learned that a pattern needed to have six squares (one for each side, or face, of the cube). We also decided that in order for a pattern to be accepted as fair, each square in it had to have one edge in common with at least one other square in the pattern. (In other words, just having a vertex in common was not enough.) During the lesson, children learned about patterns that were rotations or reflections of other patterns. And eventually (in both classes) they were able to come up with all eleven different patterns, given below. They were captivated by the lesson. And at the end, each child had a collection of cubic inches to take home.

Given paper with a one-inch square grid, make a cube 1 by 1 by 1 inch.

(1) How many different "plans" are there for this cube?
(2) How do you know that you did not miss some plans?
(3) Explain when two plans are different.

Answers:

(1) Illustration 82 shows all the plans. Each square represents one square inch.
(2) Proof by cases.
   Proof by cases usually requires classifications. (It is interesting how people classify "unfamiliar" geometric shapes.) One way of classification:
   The longest possible straight chain is 4 squares, so here are all possible cases.
   Chains:
   - Case 1. 1 four-chain, 1 three-chain, (plans 1, 2)
   - Case 2. 1 four-chain, no three-chain, (plans 3, 4, 6, 10)
   - Case 3. 2 three-chains, (plans 5, 11)

ILLUSTRATION 82. All eleven plans that will fold into a cube.

212

- Case 4. Only one three-chain, (plans 7, 9)
- Case 5. No three-chains, (plan 8)

In each case, exhausting the search for all plans is feasible. Other possible classifications can be based on other configurations, such as:

L: M   or   T: M M M   or   S: M M
M   M   M M
M M

(Here, the plans are presented in a symbolic way; each M represents one square inch.)

(3) Probably the best explanation that two plans are different is by transformation. When I cut out two plans, they are the same if one can cover the other one exactly. (This is a simple test.)

Comment: This is a nice way to introduce the concept of orientation. Let students make the cutouts from paper that has different colors (or patterns) on both sides (origami paper or paper with dots on one side only), and see if they always put one shape on a similar one, preserving the color or the pattern (red on red, but not red on white; or dots on dots, but not dots on white). This is also a nice beginning for studying geometric symmetries.

CHAPTER 51

Making Open Cubic Boxes

PROPS

Sheets of rather stiff paper from a one-inch dot grid (see handout at end of book), cut into pieces of size 3 inches by 5 inches, and a few in sizes 2 inches by 5 inches; rulers, scissors, cellophane tape. Children should also get a lot of extra paper for sketching.

THE LESSON

(1) Forming Teams

Children are divided mainly into teams of 3. In addition, one, two or three teams of 2 should be formed. Division into teams is already a part of the lesson.

(2) Explanation of the Task to Be Done

Each child is going to make an open-topped box, 1 inch by 1 inch by 1 inch, by cutting a pattern of 5 connected squares, creasing it and bending along the creases, and putting the box together using cellophane tape. Teams should propose and draw on the blackboard as many ways of forming patterns as possible.

(3) Real Task

Each team of 3 gets only one sheet of stiff paper, 3 inches by 5 inches; each team of 2 gets a 2 inch by 5 inch sheet. They have to figure out how to divide their sheet into 3 patterns for boxes (in the case of the 3 inch by 5 inch sheet) or 2 patterns for boxes (in the case of the 2 inch by 5 inch sheet). Proposals are discussed and drawn on the blackboard.

Actual Work

Children cut the sheets (they may decorate them using crayons or markers) and make the boxes.

(4) Discussion

Question: What other sizes of sheets can be partitioned into patterns for boxes without any waste? (In early grades, look only into some specific cases, 4 by 5, 4 by 3, and so on.)

Remarks

- Even for an advanced group of children, covering the whole topic requires more than two class periods.
- Parts 1, 2, and 3 can be done in early elementary grades, with the stress on skills of drawing and cutting. The teacher may have to take an active role in helping children to design patterns.
- Part 4 is far more advanced mathematically and probably cannot be used before fourth grade.

Solutions

(1) Every number greater than 1 is a sum of 3's and 2's. In addition, we may convert 2's into 3's and back according to the rule,

$$2 + 2 + 2 = 3 + 3$$

Number	Sum of 2's and 3's	Number of Teams of 2	Number of Teams of 3
2	= 2	1	0
3	= 3	0	1
4	= 2 + 2	2	0
5	= 2 + 3	1	1
6	= 2 + 2 + 2	3	0
	= 3 + 3	0	2
7	= 2 + 2 + 3	2	1
8	= 2 + 2 + 2 + 2	4	0
	= 2 + 3 + 3	1	2
9	= 2 + 2 + 2 + 3	3	1
	= 3 + 3 + 3	0	3

Remark: In higher grades, when children know some algebra, this problem can be reformulated as follows:

- number of people n
- number of 2-teams x
- number of 3-teams y

We are solving (in whole numbers only) the equation

$$2*x + 3*y = n.$$

(2) Illustration 83 shows patterns for open boxes. B is the bottom, blank squares are sides.

(3) Cutting the rectangles:

Three by five.

X X X X Z
X Y Z Z Z
Y Y Y Y Z

Two by five.

X X X X Y
X Y Y Y Y

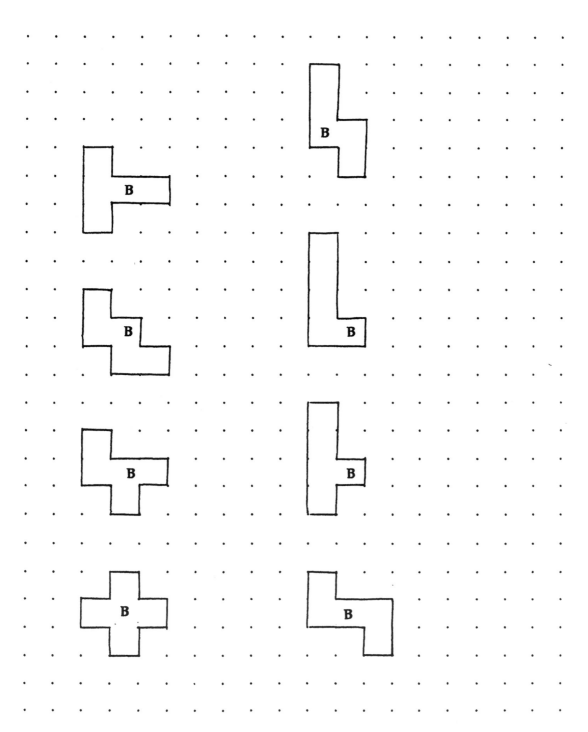

ILLUSTRATION 83. All eight patterns for open cubic boxes. B indicates the bottom of the box.

(4) Looking at the general case.

For higher grades:

(a) Because each pattern consists of 1 by 1 inch squares, a rectangle that can be divided into patterns with no waste has sides whose lengths are whole numbers (of inches).

(b) The area of such a rectangle is divisible by 5, because each pattern has an area of 5 square inches. Because 5 is prime, one of the sides also must be divisible by 5.

Thus we see that only those rectangles that have whole numbers (greater than one) for sides, and for which (at least) one side is divisible by 5, can be partitioned into patterns for boxes.

Now let's see that indeed each such rectangle can be divided into patterns.

(1) Cut the rectangle into strips that are 5 by something.

(2) From each strip, cut parts that are 5 by 2, until only a 5 by 2 or a 5 by 3 part is left.

(3) Split each 5 by 2 part, and each 5 by 3 part, as shown above.

Notice that in cutting the strip, we use the same principle as for dividing the class into teams of two and three.

For lower grades:

Look at specific cases.

Examples:

- With a 3 by 4 rectangle, 2 squares must be wasted because $12 = 5 + 5 + 2$.
- With a 5 by 5 rectangle, split it into a 2 by 5 rectangle and a 3 by 5 rectangle, and we know what to do, from above.

Do not insist on any general formulation of the method. Young children are better at knowing and doing than at explaining what they know and what they do.

CHAPTER 52

Cubic Inches

PREREQUISITE

The lesson "Square inches" (in *Breaking away from the Math Book: Creative Projects for Grades K-6*). Children should also know how to draw a plan and make a cube from construction paper (see Chapter 50, Making a cube).

THE LESSON

Each child makes 5 cubes from construction paper, having volumes 1, 2, 4, 6, and 8 cubic inches (see Illustration 84). Each child should have a paper or plastic bag to keep the cubes in.

(1) Children make a 1 cu. in. cube and a 2 by 2 by 2 inch cube. Stacking 8 small cubes together, they see that the volume of the big cube is 8 cu. in.

(2) Question: Given the length of an edge of a cube, what is its volume?

Edge	Volume
1 in.	1 cu. in.
2 in.	8 cu. in.
3 in.	27 cu. in.

We see that

$$1*1*1 = 1$$
$$2*2*2 = 8$$
$$3*3*3 = 27$$

This is a general rule: In order to compute the volume of a cube, you may multiply three copies of the length of an edge. (Do not forget to write the units correctly!) *Remark:* The rule should be explicitly stated. It is incorrect to indicate that the rule follows from these examples and force children to "discover" it.

Multiplying three copies of a number is called computing the cube of the number.

219

ILLUSTRATION 84. Five cubes, having volumes 1, 2, 4, 6, and 8 cubic inches.

(3) How to compute the cube of a number on a calculator?
The program [*][=][=] does it, for example,

Program	Display
[2][*][=][=]	8.
[3][*][=][=]	9.
[2.5][*][=][=]	15.625
[12.3756][*][=][=]	1895.3908

Remark: The last example shows that this program is better than repeated multiplication, which would be [12.3756][*][12.3756][*][12.3756][=], which is longer and is error prone.

(4) How to find a number whose cube is approximately 4? In order to make a cube which has a volume of 4 cu. in. we need to compute the length of its edge. So we have to find a number whose cube is 4.

(1) Trial and error method: 1 is too small, and 2 is too big.
Try 1.4:
[1.4][*][=][=] gives 2.744, too small.
Try 1.8:
[1.8][*][=][=] gives 5.832, too big.
Try 1.5:
[1.6][*][=][=] gives 4.096, a little too big.
Try 1.59:
[1.59][*][=][=] gives 4.019679, still too big.

Try 1.58:
[1.58][*][=][=] gives 3.944312, a little too small.

Do we need better accuracy? We know that the true value is somewhere between 1.58 in. and 1.59 in. So we know it with an accuracy of .01 of an inch. With rulers we can make measurements with an accuracy not better than 1/2 of 1/16 in. (half of the smallest division on a ruler) which is 0.03125 in., and 0.01 < 0.03, so we do not need any better accuracy. .59*16 = 9.44, so the edge should be a little longer than 1 and 9/16 in.

(2) A better method: The following program computes the number whose cube is y.

[y][*][√]

Now repeat [=][√][√]

An example:

Keystrokes	Display
[4][*][√]	2.
[=][√][√]	1.6817928
[=][√][√]	1.6104902
[=][√][√]	1.5931421
[=][√][√]	1.5888343
[=][√][√]	1.5877592
[=][√][√]	1.5874905
[=][√][√]	1.5874233
[=][√][√]	1.5874065

Let's save it and see the cube,

[M+][*][=][=]	4.000041	almost a perfect 4.
[MRC][MRC]	1.5874065	

Final step, conversion to 16ths of an inch,

[−][1][*][16][=]   9.398504

So the length of the edge is a little more than 1 and 9/16 in.

(5) Children compute the lengths of the edges for the remaining two cubes by any method given above, and then they make the cubes.

Remarks: Precision in construction and in measurements is important in this lesson. The cubes should look nice, so use colored paper.

CHAPTER 53

Dry Measure

This lesson was taught in a fourth grade class.

PROPS AND TOOLS

Construction paper (light-weight poster board is best), rulers, compasses, scissors, cellophane tape, dry uncooked rice.

Children work in groups of about four.

THE LESSON

Task 1

Each group makes a set of containers from construction paper (see Illustration 85).

(1) A cube (one cubic inch)
(2) A cylinder (height = 1 in., diameter = 1 in.)
(3) A cylinder (height = 2 in., diameter = 1 in.)
(4) A cylinder (height = 1 in., diameter = 2 in.)
(5) A cube (edge = 2 in.)
(6) A container with a triangular base (optional)

Task 2

Each group gets a cup of dry uncooked rice. Compare the amounts of rice that the containers hold. Write down the results.

EXAMPLE OF RESULTS

- The big cube holds 8 cubic inches.
- The flat cylinder holds four times as much as the small one.
- The tall cylinder holds twice as much as the small one.
- The small cylinder holds approximately 3/4 cubic inch.

ILLUSTRATION 85. Three cylinders and two cubes filled with rice. How do they compare in volume?

Task 3

Children (and the teacher) discuss the results and ask questions, for example,

- What is the volume of a cylinder that has a height of 2 in. and a diameter of 2 in.? Answer: 8 times the volume of the small cylinder, or 6 cubic inches.
- What is the volume of a box (1 in. by 2 in. by 3 in.)? Answer: 6 cubic inches.

REMARKS

Children may want to answer the questions experimentally, by making new containers and measuring. This should go on until the children tire of it. Using colorful construction paper makes the task more interesting. Precision in making containers is very important. Children need to be shown how to fill containers without making a "heap."

CHAPTER 54

Boxes for Chocolate

This lesson is suitable for upper elementary and middle school students.

INTRODUCTION

Gifts are traditionally packaged in attractive containers, and chocolate is no exception to this rule. Tobler puts some of its products in boxes with triangular cross sections. Cross sections of some boxes for Droste chocolate are hexagonal (see Illustration 86).

PROPS

Rulers, compasses, protractors (optional), scissors, glue or scotch tape, assorted construction paper and poster board, crayons or colored markers.

THE LESSON

Task

- Design, construct and decorate two boxes. The first box is 15 cm long. Its cross section is an equilateral triangle having a perimeter 12 cm long. The second box is also 15 cm long. Its cross section is a regular hexagon having a perimeter 12 cm long (see Illustration 86.)
- Compare the volumes and surface areas of both boxes.

GEOMETRICAL HINTS

Use a protractor or compass for constructing an equilateral triangle, and use a compass for constructing a regular hexagon. Draw detailed plans on plain paper before you move to poster board or construction paper. If you are using crayons or paint, make your design and color it before you glue or tape your box together.

Calculations (see Illustration 87.)

A side of the hexagon is $x = 12/6 = 2$ cm. A side of the triangle is $2x = 12/3 = 4$ cm. Call the length of both boxes L = 15 cm.

ILLUSTRATION 86. Boxes for chocolate, with triangluar and hexagonal cross sections.

Consider a smaller equilateral triangle with side $x = 2$ cm. Call its area A. Notice that the area of the bigger triangle is 4A, and the area of the hexagon is 6A. (Do you see it?)

For each box, the volume equals the area of the cross section times the length, and the area of the surface is the perimeter of the cross section times the length, plus twice the area of the cross section.

The area A of the small triangle is

$$A = \frac{1}{2} * \text{base} * \text{height}$$
$$= \frac{1}{2} * 2 * \sqrt{3}$$
$$= \sqrt{3} = 1.73 \text{ sq. cm.}$$

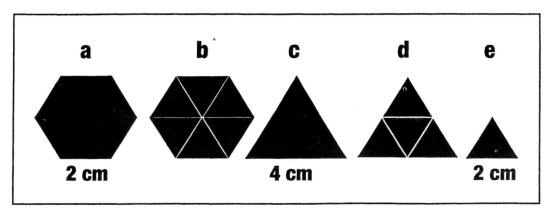

ILLUSTRATION 87. The hexagon and big triangle are made up of smaller triangles. The small triangle (e) has side 2 cm. Call its area A. The big triangle (c and d) has side 4 cm and area 4A. The hexagon (a and b) has side 2 cm and area 6A.

[The height of an equilateral triangle with base 2 is $\sqrt{3}$, by the Pythagorean theorem (see Chapter 57).]

Formulas for the triangular box:

Volume = 4*A*L = 104 cu. cm
Area = 6*x*L + 8*A = 194 sq. cm

Formulas for the hexagonal box.

Volume = 6*A*L = 156 cu. cm
Area = 6*x*L + 12*A = 201 sq. cm

Remarks: The actual sizes of the boxes will vary. Areas and volumes of your boxes can differ by a few square and cubic centimeters from the values computed above.

Comparison of the Volumes and Surface Areas

When we are making comparisons we can ask two different questions: How much more (what is the difference)? Or, how many times more (what is the ratio)? There is no mathematical rule about asking questions. They depend on what we want to know.

What are the differences?

The volume of the second box (with the hexagonal cross section) is 52 cu. cm bigger than the volume of the first one (with the triangular cross section). The surface area of the second box is 7 sq. cm larger than the the surface area of the first one.

What are the ratios?

The volume of the hexagonal box is 1.5 times bigger than the volume of the triangular one. You may also say it differently: The volume of the hexagonal box is 50% bigger than the volume of the triangular one. The surface area of the hexagonal box is 1.03 times larger than the surface area of the triangular box. Or you may say that its area is 3% larger.

CHAPTER 55

Solids Built from Equilateral Triangles

Making the polyhedra in this lesson is suitable for grade five through middle school. The activities in Part B are even appropriate for high school.

THE LESSON

Part A: A Grid of Equilateral Triangles and Some Simple Three Dimensional Objects

One of the easiest pictures that can be drawn with a compass and ruler is a grid of equilateral triangles. Simply draw circles, all with the same radius. The second circle should have its center somewhere on the circumference of the first one. After that, the point of the compass should be placed where two circles intersect. Continue drawing circles until you have a large number of them. Then connect the points of intersection with straight lines (see Illustration 88).

From equilateral triangles we can construct an unlimited variety of three-dimensional objects, some having interesting shapes. Illustration 89 gives plans for an octahedron with two windows and a pyramid without a bottom square. Illustration 90 shows the completed objects.

ACTIVITIES

Children need paper (light-weight poster board is best), pencils, compasses, rulers, and cellophane tape.

(1) **Activity 1:** Children learn to use a compass and ruler to create a grid of equilateral triangles such as the one shown in the first illustration. Precision is important, so this task should be done slowly and more than once. After drawing a grid, children should color it with crayons or paints, creating an artistic design.

(2) **Activity 2:** Children draw a grid again, but this time they cut out from it the shapes shown in Illustration 89, crease and fold them, and tape them together to create the objects shown in Illustration 90.
Remark: Drawing the grid is an essential part of this lesson, so ready-made grids should not be used.

(3) **Activity 3:** Children should make new shapes of their choice. They may color them before taping them together. An investigation of different possible solids is an important part of this activity.

ILLUSTRATION 88. A grid of equilateral triangles.

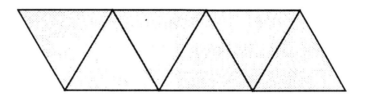

octahedron with two windows

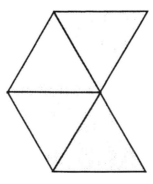

pyramid without a bottom square

ILLUSTRATION 89. Plans for an octahedron with two windows and a pyramid without a bottom square.

ILLUSTRATION 90. An octahedron with two windows and a pyramid without a bottom square. Their plans are shown in Illustration 89.

ILLUSTRATION 91. The eight convex solids made from equilateral triangles. Their plans are given in Illustration 92.

231

Remark: Creating new solids, and drawing plans for them is far more difficult than it seems to be. Therefore, especially at the beginning, hints, suggestions, and help from the teacher are needed. Also, working in groups formed from children with different levels of skill and dexterity is helpful. The teacher should have several finished objects ready, to show children what the goal of this activity is.

Part B: Convex Solids Made from Equilateral Triangles

TASK

Construct the eight convex polyhedra described below. (All their faces are equilateral triangles.) Three of them are Platonic solids (tetrahedron, octahedron, icosahedron). Illustration 91 shows the eight finished objects; Illustration 92 gives plans for making them; and Illustration 93 shows how the five which are not Platonic solids fit together.

Here is a description of the eight polyhedra, numbered as in Illustrations 91 and 92:

(1) The simplest one is the tetrahedron. It has 4 faces, 4 corners, also called vertices, and 6 edges.

The next three are double pyramids.

(2) Take two pyramids with square bases and glue the bases together. This polyhedron is called the octahedron. It has 8 faces, 6 corners, and 12 edges.
(3) Take two tetrahedra (they are pyramids with triangular bases), and glue them together. This polyhedron has 6 faces, 5 corners, and 9 edges.
(4) Finally take two pyramids which have pentagonal bases and glue them together. This polyhedron has 10 faces, 7 corners, and 15 edges.

In order to describe the next two, look how to build a "fences" from the triangles (see Illustration 94). Now, if we crease the strips above and glue the left edge to the right edge, we get "fences." The upper one has a pentagon at the top and at the bottom. The lower one has a square at the top and at the bottom.

(5) Construct the first fence. Add to it two pyramids with pentagonal bases, one to the top, and one to the bottom. You get a polyhedron called an icosahedron. It has 20 faces, 12 corners, and 30 edges.
(6) Add to the second fence two pyramids with square bases. You get a polyhedron with 16 faces, 10 corners, and 24 edges.
(7) Think about a prism (a roof) consisting of three squares and 2 triangles (see Illustration 95). Now add a square based pyramid to each side of the roof and to the bottom square (3 pyramids in all). You create a polyhedron with 14 faces, 9 corners, and 21 edges.
(8) Finally, look at an octahedron with two (touching) triangles missing. It looks like a bird (or dinosaur) head with open beak. Take two of them and put them beak to beak, turning one 90 degrees relative to the other. You get a polyhedron with 12 faces, 8 corners, and 18 edges.

MAKING POLYHEDRA

This task should take several class periods. Children should not construct more than two simple polyhedra or one complex one during a day. After this unit is finished, each child should have a complete set of eight polyhedra.

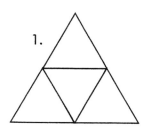

1.

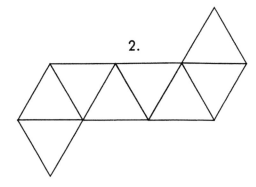

2.

3.

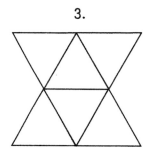

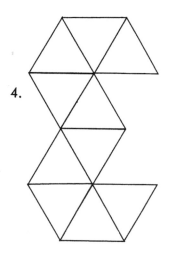

4.

5.

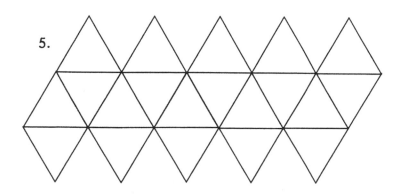

ILLUSTRATION 92. Plans for making the eight convex solids made from equilateral triangles.

233

6.

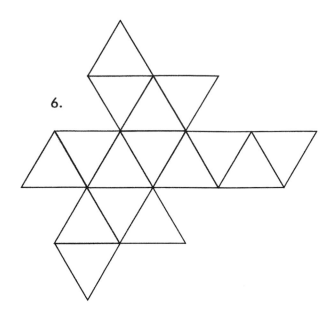

7.

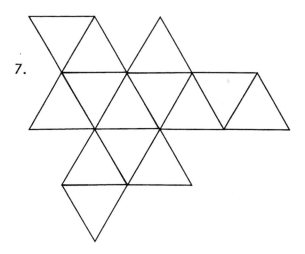

8.

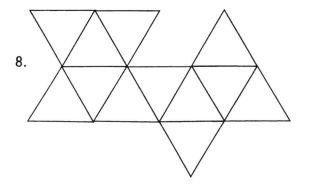

ILLUSTRATION 92. Continued.

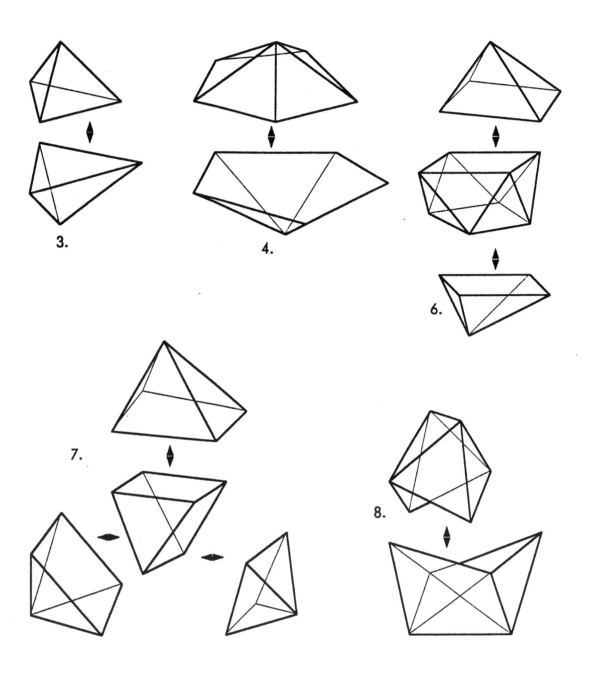

ILLUSTRATION 93. How the five deltahedra that are not Platonic solids fit together. The numbers match those in Illustrations 91 and 92.

235

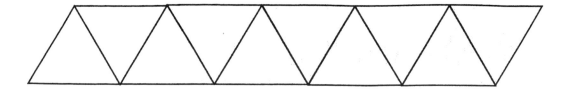

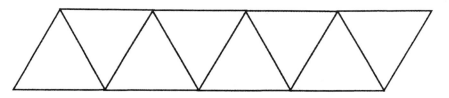

ILLUSTRATION 94. "Fences" made from triangles.

MATERIALS

Strong poster board in bright colors should be used. It is better to use transparent cellophane tape, instead of glue. Tape is easier to handle and less messy.

TOOLS AND SKILLS

Good compasses, good scissors and straight rulers are essential. Children must be shown how to make a pattern of equilateral triangles with a compass and a ruler (see Part A), how to cut them out precisely, how to crease them to create smooth and straight edges, and how to fix them with cellophane tape. This is a long unit and by doing all the tasks children should acquire considerable manual skills.

A SCHEMA FOR A LESSON

- The teacher starts with a verbal description of a polyhedron, and by showing a finished model.
- Next, patterns for the models are proposed and discussed. (Different children may use different patterns.)
- Patterns are drawn and cut out, and polyhedra are constructed.

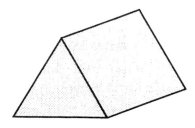

ILLUSTRATION 95. A prism consisting of three squares and two triangles.

During the first few sessions the teacher should do most of the planning, because children will struggle with the technical details of drawing, creasing, and cutting. As their skills improve, they should take over the design. The children may help each other; this is especially needed during the assembly phase. One person holds a polyhedron, and the other fastens it with transparent tape.

OTHER ACTIVITIES WITH FINISHED POLYHEDRA

Regular Polyhedra and Euler's Theorem

Only three of these eight polyhedra—the tetrahedron, octahedron, and icosahedron—seem to have generally accepted names. For the other five, children may be asked to create their own names. The names must be agreed upon and then they should be used in discussion of these solids in class.

The tetrahedron, octahedron, and icosahedron are called regular because (1) all their faces are congruent (equilateral triangles of the same size); (2) the same number of edges meet at each corner.

Observe that the number of edges meeting at one corner is 3, 4, or 5. (Six triangles could not form a corner; their angles add up to 360 degrees.)

Children should create a table to compare these polyhedra, for example,

Which one?	Number of			Number of corners with		
	Faces	Corners	Edges	3 Edges	4 Edges	5 Edges
(1) tetrahedron	4	4	6	4	0	0
(2) octahedron	8	6	12	0	6	0
(3)	6	5	9	2	3	0
(4)	10	7	15	0	5	2
(5) icosahedron	20	12	30	0	0	12
(6)	16	10	24	0	2	8
(7)	14	9	21	0	3	6
(8)	12	8	18	0	4	4

Observe some numerical properties:

(1) faces + corners − edges = 2.
   This is called Euler's formula. This formula is true for all polyhedra.
(2) (3/2)*faces = edges. This is true for all polyhedra with triangular faces.

   Proof.
   Each face has 3 edges. But if we count 3*faces, each edge is counted twice because it belongs to 2 faces. So the correct number of edges is 3*faces/2.

   Question: If you combine these two formulas, you can find the relation between the number of faces and the number of corners for polyhedra with triangular faces. What is the relation?
   Answer:

$$faces + corners - (3/2)*faces = 2.$$

Therefore,
$$corners = (1/2)*faces + 2$$

Similarly, the relation between number of corners and number of edges is
edges = 3*corners − 6.

REMARK

As mentioned, this unit may be used also in middle or high school. It could be used as a preliminary activity for computing volumes of polyhedra, or for studying groups of symmetries of polyhedra, or for studying movement in space.

CHAPTER 56

Three Congruent Pyramids That Form a Cube

This lesson was taught in a sixth grade class.

THE LESSON

A cube can be divided into three congruent pyramids with square bases. Construct these three pyramids from paper or poster board and assemble them into a cube (see Illustrations 96–98).

Method

Draw the plan. The base is a square. All four sides are right triangles (see Illustration 97). If the side of a triangle is a unit, then $x = \sqrt{2}$ and $y = \sqrt{3}$. We used 5 cm as the length of the side of our square. Then $x = 5\sqrt{2}$ and $y = 5\sqrt{3}$ cm.

All three pyramids are the same. Cut them out, assemble them, and form a cube.

Notice that 24 such pyramids form a larger cube! Constructing it is a beautiful class project.

ILLUSTRATION 96. Three congruent pyramids that form a cube.

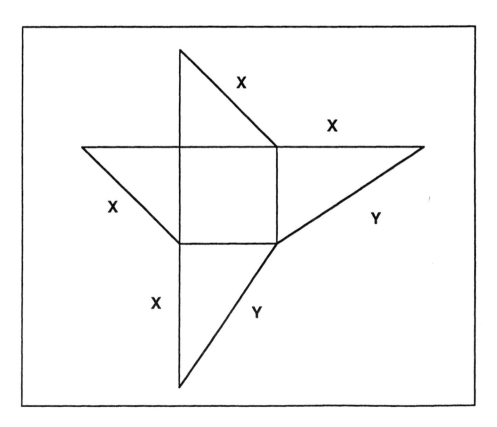

ILLUSTRATION 97. A plan for a pyramid.

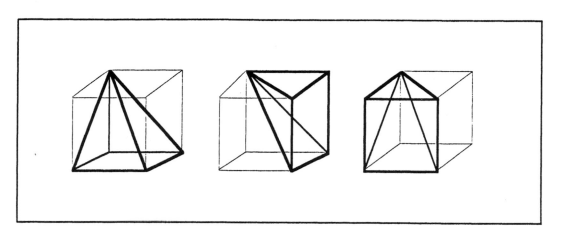

ILLUSTRATION 98. How the three pyramids fit together to form a cube.

PYTHAGOREAN THEOREM

CHAPTER 57

Pythagorean Theorem

INTRODUCTION

The Pythagorean theorem states that if you build three squares on the three sides of a right triangle, the sum of the areas of the squares built on the legs is equal to the area of the square built on the hypotenuse (see Illustration 99).

THE LESSON

A modern version of this theorem, which has many applications, says: The length of the hypotenuse is the square root of the sum of the squares of the lengths of the legs, $c = \sqrt{(a^2 + b^2)}$ (Illustration 100).

A program that computes c (clear the memory):

$[a][*][M+][b][*][M+]\ [MRC][\sqrt{}]$.

The theorem is named after Pythagoras (about 572–497 B.C.), a Greek philosopher and mathematician, but there is no historical evidence of Pythagoras' individual mathematical achievements [see Heath, T. *A History of Greek Mathematics, Vol. 1. From Thales to Euclid.* New York: Dover (1981; 1921)]. The theorem was discovered several times in different places, so it is impossible to attribute it to one person.

One way to look at the Pythagorean theorem is to consider the following task: Given two squares (with sides a and b), cut them into pieces, and from these pieces build one big square. We will see that the big square will have a side of length c, which is the hypotenuse of a right triangle with legs a and b (look again at Illustrations 99 and 100). So by doing the task we will prove the Pythagorean theorem.

Task 1 (Easy)

Given two squares of the same size, cut them into parts and build one big square (using all the parts). You will need two squares of origami paper of the same size but different colors, scissors, and an envelope (one per child).

There are many ways to do it. The simplest one is to cut both squares along their diagonals and arrange the four pieces into a square. (All other methods require at least five pieces.) The envelope is for keeping all parts for future use (see Illustration 101).

245

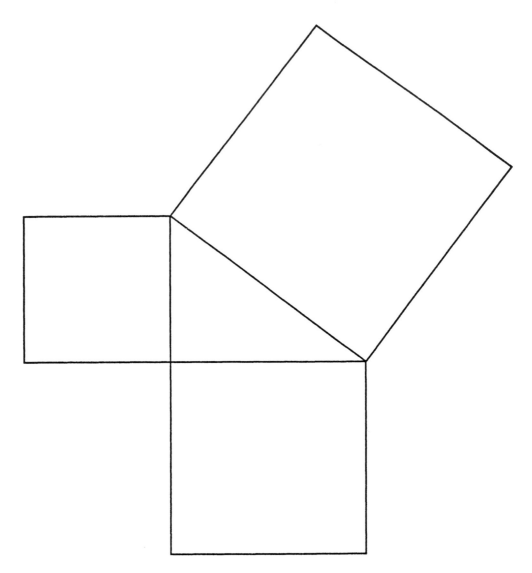

ILLUSTRATION 99. The sum of the areas of the smaller squares equals the area of the big square.

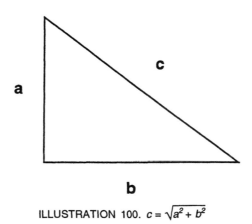

ILLUSTRATION 100. $c = \sqrt{a^2 + b^2}$

246

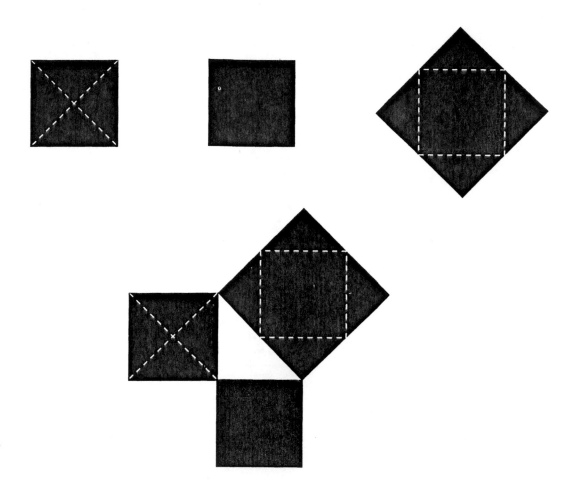

ILLUSTRATION 101. One way to cut two squares of the same size and build one big square using all the parts. (Is there another way?)

Task 2

Given two squares of different sizes, cut them into parts and build one big square (using all the parts). You will need two squares of origami paper of different sizes and different colors, scissors, ruler, pencil, protractor (optional), and an envelope (one per child).

Again, there are many ways of doing the task. The method below uses the smallest number of parts (five) and clearly shows why the Pythagorean theorem is true.

Given two squares, put one at the top of the other and draw line c as shown in Illustration 102. Now move the smaller square to the right and down (Illustration 103), and draw a straight line from X to Z (Illustration 104).

Remark: Already you may notice that segment *XY* has length a. (You took away b from the left, and added b to the right.) So the right triangle *XYZ* has sides a, b, and c. (In the next step, after you cut, check your conclusion by putting one triangle on top of the other.) Also, angle *TXZ* is a right angle (measure it with a protractor).

Cut along line c, and along line *XZ*. Now the big square is divided into three parts,

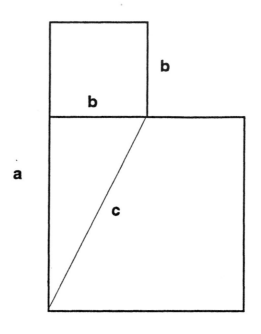

ILLUSTRATION 102.

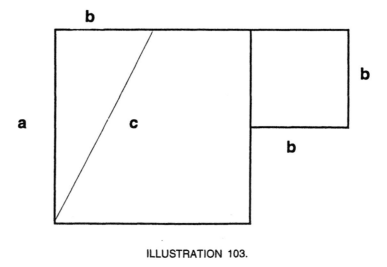

ILLUSTRATION 103.

248

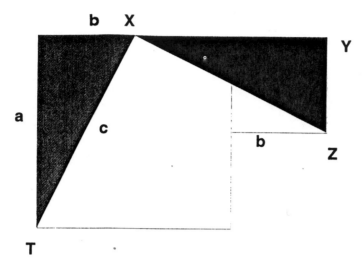

ILLUSTRATION 104.

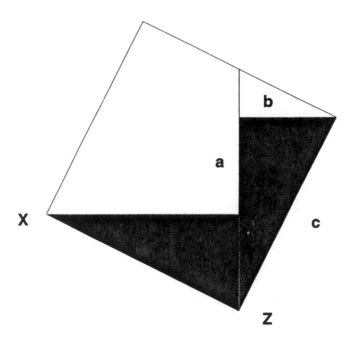

ILLUSTRATION 105.

and the small one into two parts. Now move triangle *XYZ* down, and triangle *abc* to the right (see Illustration 105).

So we have a new square whose area is the sum of the areas of the two smaller ones and whose side is the hypotenuse of a right triangle with sides *a* and *b*. So we have proved the Pythagorean theorem!

Final Remarks

There is a lot of confusion concerning mathematical proofs, especially in geometry. What is given above is an informal proof, namely a reasoning which, by any means, establishes the truth of some mathematical statement. There is another type of proof (sometimes called a demonstration, or a derivation) which shows that some mathematical statements follow logically from some other mathematical statements (often called axioms). Because this second type of proof was first used in geometry, and was not generally used in other mathematical theories until the nineteenth century, an impression was formed that demonstrations are specific to geometry. This is not so. At present all mathematical theories have their axioms, and in research papers in mathematics demonstrations are obligatory. On the other hand, informal proofs in geometry, based on a drawing, symmetry, or cutting and rearranging geometrical figures, are often interesting, convincing, and easy to understand (even if they are sometimes very difficult to find). Without taking a stand regarding at what age children should start reasoning about mathematical objects, we think that informal geometrical reasonings are probably the best introduction to the concept of a proof, and geometric demonstrations (reasoning based on Euclid's axioms) are the worst.

PYTHAGOREAN PUZZLE

This lesson can be embedded in a story about selling land, planting grass, cutting cloth, or many others. It can also be presented in higher grades (e.g., middle school) as a proof of the Pythagorean theorem, because it contains such a proof. Here the lesson is provided without any embelishments.

Preparation

Children work in groups of four. Each group will make four identical right triangles from construction paper. (Different groups may use paper of different colors, and make triangles of different shapes and sizes.) The legs of the triangles should be measured and written down, For example:

Legs: $a = 6.5$ cm, $b = 12.3$ cm

Then children should draw one square whose sides are $a + b = 18.8$ cm long.

Problem

(1) Put the four triangles on the square (in the four corners) as shown in Illustration 106. What is left uncovered is one square whose sides have a length (measure it) of $c = 13.9$ cm. (The side is the hypotenuse of the right triangle.)
(2) Now put all four triangles in two corners of the square as shown in Illustration 107. What is left uncovered are two squares whose sides have lengths a and b.

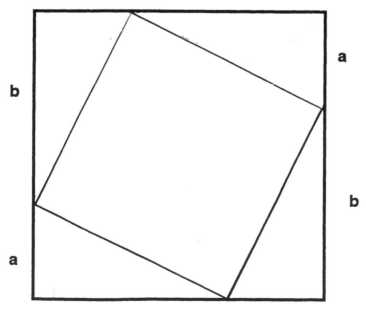

ILLUSTRATION 106.

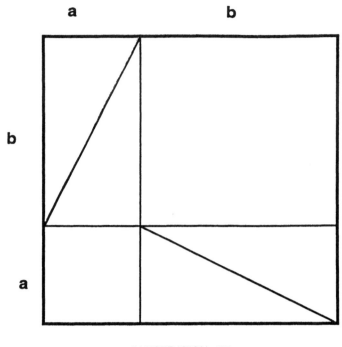

ILLUSTRATION 107.

In which of the two instances was a bigger area left uncovered? Let children make both coverings as many times as they want, and do not press them too soon for an answer. The answer is, the area uncovered is *the same,* because each time we covered the same area of a big square using the same four triangles.

Remark: It is likely that some children will not "get it." This is not necessarily surprising, because the principle involved is: If, from a region of area x, some parts of total area y are removed, what remains has an area of $x - y$, independent of its shape. This is not so obvious because "how big it looks" depends very much on its shape.

YET ANOTHER PROOF OF THE PYTHAGOREAN THEOREM

Draw two identical right triangles, and label their legs a and b and their hypotenuse c. Make the hypotenuse c be the base of each of the triangles, and construct an altitude for each from the base up to the right angle. The altitude divides the base into two segments. Label them x and y. Then $x + y = c$ (see Illustration 108). Cut one of the triangles along its altitude. You now have three triangles, which are similar. Thus $b/c = y/b$ and $a/c = x/a$, so $b^2 = cy$ and $a^2 = cx$ So $a^2 + b^2 = cx + cy = c(x + y)$.

But $x + y = c$, so $a^2 + b^2 = c^2$.

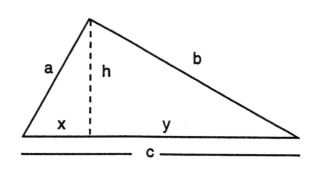

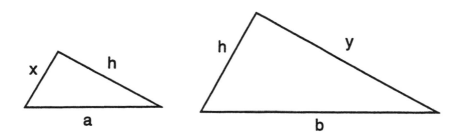

ILLUSTRATION 108. Three similar triangles.

PYTHAGOREAN TRIPLES

Pythagorean triples are three different whole numbers a, b, and c, such that $a^2 + b^2 = c^2$. Notice that if you have one triple, you can multiply the three numbers by the same whole number and get another triple. For example, one triple is 3, 4, 5. Multiplying by 12, another is 36, 48, 60.

You can get *all* Pythagorean triples as follows. For all positive integers x and y, with $x > y$, a triple is obtained by $x^2 - y^2$, $2xy$, and $x^2 + y^2$. Below we have begun a table, with $x > y$. Continue it and see what Pythagorean triples you can find.

x	>	y	$x^2 - y^2$	$2xy$	$x^2 + y^2$
2		1	3	4	5
3		1			
3		2			
4		1			
4		3			
5		1			
5		2			
5		3			
5		4			
6		1			
6		5			
7		1			
7		2			
7		3			
7		4			
7		5			
7		6			
8		1			
8		3			
8		5			
8		7			
9		1			
.		.			
.		.			
.		.			

MISCELLANEOUS

CHAPTER 58

One Third

Children (and adults), when presented with something new, try to make sense of it in their own way, which is often in disagreement with accepted explanations. Young children (kindergarten and first grade), especially if they use calculators, are exposed to decimal fractions long before they learn and understand the logic of decimal notation.

That 0.25 is a quarter is not a problem; 25 cents is a quarter of a dollar. There is some problem with 0.5 because, "It should be 0.50, because 50 cents is one half of a dollar." But even in these examples one should be aware that most children do not see the decimal point as marking a division between the whole and the fractional part of a number. They view it as separator between two whole numbers, dollars and cents. They often write .25¢ or $2.50¢, "two dollars and fifty cents."

Of course the biggest problem is with 1/3. Why does the calculator show 0.3333333? "Three is because of one third, but why so many threes?" Saying that "the calculator just does it," is not good enough. Children need a reason. One child explained 0.5 as follows, "The calculator knows that 5 dimes is half a dollar. But it does not know that it is 50 cents. So it shows .5 for five dimes." A child, asked what she thought that 0.3333333 means, answered, "3/7. There are seven threes."

THE LESSON

Children work in groups of three. (It may be necessary to form one group of two or four.) Each group gets one 6 oz. can of Play-Doh, or modeling clay, three small paper plates, one bigger paper plate, and rulers. (The rulers may be used for both measuring rolled clay and cutting it.)

Step 1: Making Balls

Children in each group make 10 balls of clay of equal size. They are put onto the big plate.

Remark: This is not an easy task. Children may need some help and advice. They should be given plenty of time for it. (The balls are approximately 1 inch in diameter.)

Step 2: Sharing

Children divide 9 of the balls among their three small plates, leaving one on the big plate.

ILLUSTRATION 109. Dividing clay into thirds.

Step 3: Making More Balls

Children make 10 small balls from the one remaining big ball, and put them on the big plate.

Step 4: Sharing

Children divide 9 of the small balls among their three small plates, leaving one on the big plate (see Illustration 109).

These steps can still be repeated twice. So at the end, each child has on his or her plate:

- 3 big balls
- 3 small balls
- 3 very small balls
- 3 specks of clay

And a speck is left on the big plate.

Explanation

How can you get 1/3 of a lump of clay? Divide the lump into 10 equal parts. Give each person 3. One is left. Divide it into 10 parts. Give each person 3. One is left. Do it over and over again.

On your plates, balls in different sizes come in threes.

Now, 1/3 on the calculator.

Keystrokes	Display	Comments
[1][/][10][=]	0.1	Divide one into 10 parts.
[*][3][=]	0.3	Three parts for each person.
[0.1]	0.1	One part is left.
[/][10][=]	0.01	Divide what is left into 10 parts.
[*][3][=]	0.03	Three parts for each person.
[0.01]	0.01	One part is left.

(you may still continue)

How much does one person have so far?

  [0.3][+][0.03][=] returns 0.33

Can we guess how much a person will have after the next round?

Another Explanation

How is 0.3333333 made?

3	is for 3 times one tenth.
3	is for 3 times one tenth of what was left.
3	is for 3 times one tenth of what was left.
3	is for 3 times one tenth of what was left.
3	is for 3 times one tenth of what was left.
3	is for 3 times one tenth of what was left.
3	is for 3 times one tenth of what was left.

Why no more threes? It is too small even for a calculator.

Comment: Do not expect that a child will understand what is going on in a way similar to an adult. Do not overexplain. An "ideal" understanding:

- Why so many threes?
- Each one is 3/10 of what was left.

CHAPTER 59

Statistics with Pennies

This lesson is suitable for upper elementary and middle school grades.

THE LESSON

(1) Provide students with a jar of pennies, and have them make a distribution of the years the pennies were minted.
(2) Try to interpret the distribution: "What happens to a penny?"

Example

The pennies used here were collected in 1995 in Boulder, Colorado, near the Denver mint.

DISTRIBUTION

Year:	1965	66	67	68	69
Number:	0	1	0	0	2
Percent:		2.3%			

Year:	70	71	72	73	74		75	76	77	78	79
Number:	0	1	2	2	1		2	2	2	0	1
Percent:			4.6%						5.4%		

Year:	80	81	82	83	84		85	86	87	88	89
Number:	5	2	1	4	4		1	2	1	4	4
Percent:			12.3%						9.2%		

Year:	90	91	92	93		94	95	
Number:	7	6	2	6		22	43	total = 130
Percent:		16.2%				16.9%	33.1%	

Let's look at the data differently:

- 1994 and 1995 cover 50%, and their ratio is 1 to 2.
- 1985 to 93 (9 years) covers 25%.
- 1970 to 84 (15 years) covers 22%.

This is a strange distribution. It has a strong recency effect. Look at the last four years:

Year	Number of pennies	Percent
95	43	33%
94	22	17%
93	6	5%
92	2	2%
91	?	?%

We could expect a geometric decrease with a factor of 2 or 3, and it would mean that earlier dates were very rare. On the other hand, the distribution in the eighties and seventies looks almost constant, without any effect of time.

It appears that there are several independent processes operating.

(1) Most freshly minted pennies do not last long. Almost half of them are removed from circulation within a year.

(2) Some of the observed variation is also due to the fact that different numbers of pennies are minted each year.

But there must be other factors that explain why some pennies live so long.

(3) Many pennies disappear from circulation, ending in "penny jars," drawers, old purses, pockets, and so on. They can stay there for years and years, and then reappear again. Almost half of all pennies have passed through such dormancy. This process is responsible for pennies from the seventies and eighties.

(4) Some old coins start getting more valuable. It is hard to find such coins because there are many people who especially look for them and take them out of circulation. This is probably the reason that we never see very old ones. We must buy them in coin shops.

Or is there another explanation for this distribution? Or maybe, the distribution above is not typical, and doesn't represent the pennies in circulation?

Here is a new statistic, which does *not* confirm the previous one.

Year:	70	71	72	73	74		75	76	77	78	79
Number:				2				4	1	1	
Percent:			4%					11%			

Year:	80	81	82	83	84		85	86	87	88	89
Number:	3	4	4	2	1		1			1	1
Percent:			25%						5%		

Year:	90	91	92	93		94	95		total
Number:	3	4	5	2		9	9		57
Percent:		25%				16%	16%		

The difference is that the big dominance of 1995 pennies has disappeared. This could mean that the previous one was biased because the Denver mint released freshly minted pennies to local businesses.

CHAPTER 60

The Concept of Infinity

THE LESSON

How many diameters does a circle have altogether? (Infinitely many.) How many numbers are there? I want to know the total. (Infinitely many.) How many digits does the square root of two have? (Infinitely many.) If I add all the whole numbers, what is the sum? (Infinity.)

So what is an infinity? This is a complex concept, very basic, which began to be studied very recently (less than 200 years ago). One of the main findings is that we cannot talk about one infinity; instead, we talk about many different ones. Thus we do not deal with just one concept, but with several different but related concepts.

Let us look at some of the uses of these concepts.

(1) There are infinitely many whole numbers. What does it mean? Well, there is no biggest number. Whenever you have a whole number, you may add 1 to it and get something bigger. This is what it really means.
(2) The number square root of 2 ($\sqrt{2}$) has infinitely many digits. What does it mean? Let's play a game. We start with 1. You ask me, "What is the next digit?" and I'll write it down. Ready?

	1.
What is the next digit?	1.4
What is the next digit?	1.41
What is the next digit?	1.4142
What is the next digit?	1.41421
What is the next digit?	1.414213
What is the next digit?	1.4142135
What is the next digit?	1.41421356

It becomes boring, but I cannot tell you, "That is it; there are no more digits," because after each digit there is a next one. But what about numbers such as 1/4? It is equal to 0.25, and there are no more digits. Let's start our game again, this time with 0.

	0.
What is the next digit?	0.2
What is the next digit?	0.25
What is the next digit?	0.250
What is the next digit?	0.2500

It is again boring. But I can tell you, "Don't bother me anymore, there are only zeroes from now on."

(3) If I added all the whole numbers, I would get infinity. What does it mean? Here the answer is different than in the previous cases. In the previous cases there is nothing impossible in asking the next question. But now the situation is different. What if I added all the whole numbers? You cannot do it, buddy! So let's see what would happen if you try.

$$1 + 2 = 3$$
$$3 + 3 = 6$$
$$6 + 4 = 10$$
$$10 + 5 = 15$$
$$15 + 6 = 21$$

This process will never end, so you never add all the whole numbers. But because the values became bigger and bigger, I can say figuratively, "If you had finished, you would get infinity."

(4) Are all infinities equal? Only every second whole number is even (the others are odd), so we have 1/2 as many even numbers as whole numbers. True or false? Let's see.

Whole numbers:	0	1	2	3	4	5	6	7	8	9	10	11
Even numbers:	0	2	4	6	8	10	12	14	16	18	20	22

Clearly we have as many whole numbers as even numbers! Simply match each even number with its half.

Did we get a contradiction? No, but we must revise some other more basic concepts, namely part and whole. What is more true than the statement, "A part is always smaller than the whole"? It is confirmed by all everyday experiences. If I have a collection of objects and keep only part of them, I have less objects.

Well, we do not have everyday experiences with infinite collections. For infinite collections, a part may have as many elements as the whole. Thus, the even numbers form a proper part of the collection of all whole numbers, but at the same time, there are as many even numbers as whole numbers.

Can all infinite collections be matched with each other in a way similar to the way we have matched the even numbers with all the whole numbers? The answer is no. Some infinite collections have more elements than others, so their elements cannot be paired together.

Only one case is relevant for school mathematics. Consider four collections:

- points on a straight line
- real numbers
- rational numbers
- whole numbers

The first two collections can be matched together (a real number line), and the last two also can be matched together (it is not so easy). But the first two collections cannot be matched with the second two.

Matching numbers with points on the line underlies the system of coordinates (analytical geometry) and most other problems of geometrical measurements. This match requires the real numbers. (There are not enough rational numbers to match all points on the line.) This is the reason why the real numbers give a mathematical foundation for all sciences, in spite of the fact that all practical computations are done with rational numbers only.

LEARNING HOW TO USE A CALCULATOR

CHAPTER 61

A Calculator Tutorial

INTRODUCTION

All computations in this book can be done on a simple four-operation calculator (Illustration 110). The sequences of keystrokes were tested on a TI-108, so if you are using another brand, some minor adjustments may be needed.

A calculator should not replace mental arithmetic. It is a tool which, in combination with mental computation, allows anyone to perform correctly complex calculations with decimals. To became an expert user of a calculator requires almost no practice. But it requires using mathematical knowledge to plan a computation carefully, and knowing the features of the calculator in order to achieve maximum efficiency. We show how to use a calculator by a set of examples that provide simple programs solving some typical tasks.

UNDERSTANDING HOW A CALCULATOR WORKS

A four-operation calculator can store inside only three numbers at the same time. One is shown on the display, another is kept in the calculator's memory, and the third is kept in a register. In addition the calculator can "remember" one of four arithmetic operations, +, −, *, or /.

Action Achieved by Keystrokes	What Is Stored and Where			
	Display	Register	Memory	Operation
Turn the calculator on, [ON/C]	0	0	0	none
Enter the number −23.5, [23.5][+/−]	−23.5	0	0	none
Store it in memory, [M+]	−23.5	0	−23.5	none
Enter the number 14.5, [14.5]	14.5	0	−23.5	none
Choose an operation, [+]	14.5	0	−23.5	+
Enter the second number, [2.25]	2.25	14.5	−23.5	+

The calculator is ready to add. Notice that the number 14.5 was not lost, but has moved to the register.

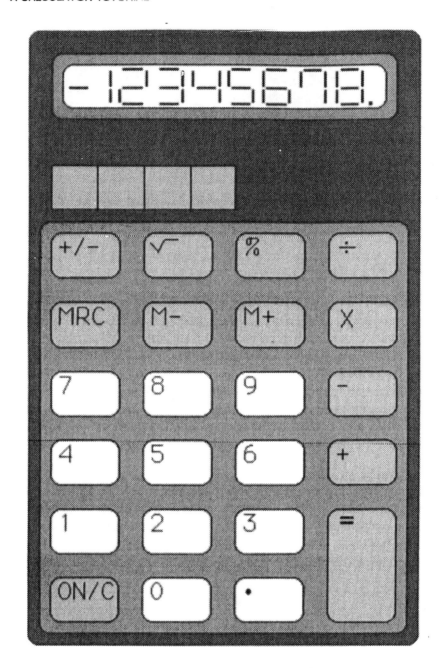

ILLUSTRATION 110.

Perform the operation,

| [=] | 16.75 | 2.25 | −23.5 | + |

Notice that 2.25 landed in the register.

Choose another operation,

| [*] | 16.75 | 2.25 | −23.5 | * |

Recall the number from memory,

| [MRC] | −23.5 | 16.75 | −23.5 | * |

16.75 has moved from the display to the register.

Perform the operation,

| [=] | −393.625 | 16.75 | −23.5 | * |

This time 16.75 in the register. To know the calculator is to know what happens, when, and why.

Clear the memory,

| [MRC][MRC] | −23.5 | 16.75 | 0 | * |

Clear the display, register, and operation,

| [ON/C][ON/C] | 0 | 0 | 0 | none |

Some operations on a calculator are achieved by pressing a sequence of keys. On simple calculators, usually you need to press at most two keys.

After a while the calculator turns itself off.

EXAMPLES OF SIMPLE PROGRAMS

In each example, start with a clear calculator. To clear everything, press [MRC][MRC][ON/C][ON/C].

(1) Add the list of numbers 123.45, 87.2, −73.8, 456.85. The first method:

Keystrokes	Display	Comments
[123.45][+]	123.45	
[87.2][+]	210.65	Do not press [=]. The next operation executes the previous one.
[73.8][+/−][+]	136.85	The change of sign to minus is done after the number is entered.
[456.85][=]	593.7	The sum is also in the memory.

A second method.

Keystrokes	Display	Comments
[123.45][M+]	123.45	
[87.2][M+]	87.2	
[73.8][M−]	73.8	Subtract 73.8.
[456.85][M+]	456.85	
[MRC]	593.7	The sum is also in the memory.

(2) Compute the sum of squares 11^2, 12^2, 13^2, 14^2. Notice first that simple calculators do *not* use precedence of operations (they do *not* do multiplication and division first, addition and subtraction, later). Thus,

$\quad$ [11][*][11][+][12][*][12][=] computes (11*11 + 12)*12 because all operations are performed in the order they are entered. Thus in this case you should keep the sum in the calculator's memory.

There is not a key for computing a square on simple calculators, but on the TI-108 and many others, the following combination,

[*][=]	computes the square of displayed number.
[*][=][=]	computes the cube.
[*][=][=][=]	computes the fourth power; and so on.

The combination,

[*][M+]	also computes the square, but it cannot be used for higher powers.

Keystrokes	Display	Comments
[11][*][M+]	121.	11^2 is in the memory.
[12][*][M+]	144.	12^2 is added to the memory.
[13][*][M+]	169.	
[14][*][M+]	196.	
[MRC]	630.	This is the required sum.

(3) Compute 1/3 + 1/5 + 1/7 + 1/11 (and give the result as a decimal). A reciprocal can be computed by the following key combinations,

    [/][=] and [/][M+].

Keystrokes	Display	Comments
[3][/][M+]	0.3333333	Decimal approximation of 1/3.
[.2][M+]	0.2	You can compute 1/5 mentally.
[7][/][M+]	0.1428571	
[11][/][M+]	0.090909	
[MRC][MRC]	0.7670994	Here is the answer. (Memory is clear.)

(4) Which is bigger, 1/7 + 1/11 or 1/6 + 1/13 ? The sign of the difference (1/7 + 1/11) − (1/6 + 1/13) tells which is bigger (a positive difference means the first is bigger, a negative difference means the second is bigger, and 0 means that they are equal).

Keystrokes	Display	Comments
[7][/][M+][11][/][M+]	0.090909	The first sum is in memory.
[6][/][M−][13][/][M−]	0.076923	The second sum is subtracted.
[MRC][MRC]	−0.0098235	The second sum is bigger, 1/7 + 1/11 < 1/6 + 1/13.

(5) Starting with 59, count down by 13. Calculators have a "constant" feature which allows you to repeat the same arithmetic operation with the same argument over and over again.

Keystrokes	Display	Comments
[59][−][13][=]	46.	13 stays in the register and minus
[=]	33	is remembered as an operation.
[=]	20	
[=]	7	
[=]	−6	Do not stop!
[=]	−19	Investigate negative numbers.
[=]	−32	
[=]	−45	
[=]	−58	

It works with addition and division in the same way. But for multiplication, the first argument goes into the register and is repeated.

Keystrokes	Display	Comments
[10][*][3][=]	30.	Multiplication is by 10 and not
[=]	300.	by 3.
[=]	3000.	

(6) Compute $\sqrt{4.32 + 1.2*2.3}$

Keystrokes	Display	Comments
[1.2][*][2.3][+]	2.76	Remember the order of operations.
[4.32][=]	7.08	
[√]	2.6608269	Square root is computed last.
[*][=]	7.0799997	Compute the square of the square root.

Because simple calculators truncate the results instead of rounding them, you get back only 7.0799997 instead of 7.08.

(7) Increase 23.86 by 12%.

Keystrokes	Display	Comments
[23.86][+][12][%]	26.7232	You do not use [=] with [%].

(8) The Fibonacci numbers are, 1 1 2 3 5 8 13 21 34 55 89. . . .

Keystrokes	Display
[1]	1.
[+][=]	1.
[+][=]	2.
[+][=]	3.
[+][=]	5.
[+][=]	8.
[+][=]	13.
[+][=]	21.

Can you figure out why this program works?

And here is another program.

Keystrokes	Display
[1][M+]	1.
[+][MRC]	1.
[M+]	2.
[+][MRC]	3.
[M+]	5.
[+][MRC]	8.
[M+]	13.
[+][MRC]	21.